SCOTT W. STARRATT

AGI Environmental Awareness Series, 2

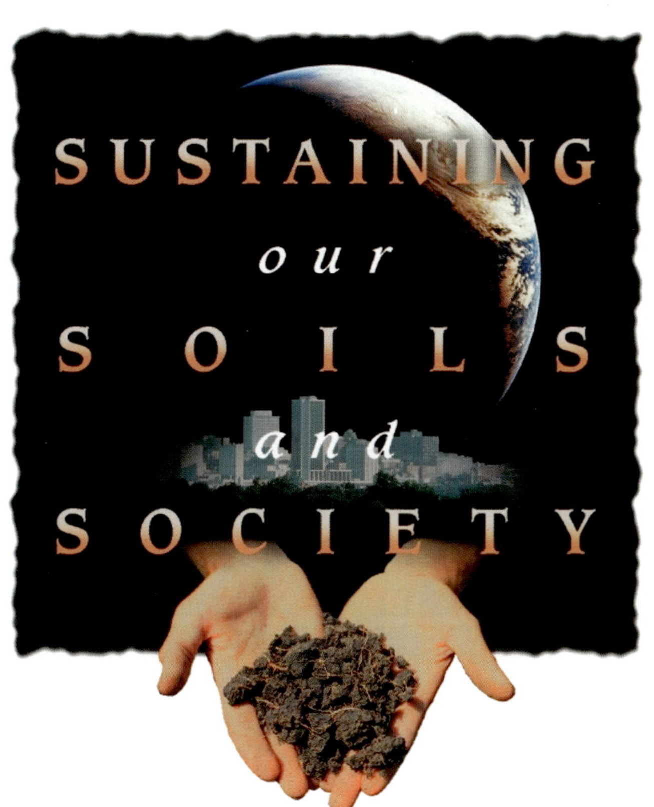

Thomas E. Loynachan
Kirk W. Brown
Terence H. Cooper
Murray H. Milford

American Geological Institute
Alexandria, Virginia

In cooperation with

American Geological Institute
4220 King Street
Alexandria, Virginia 22302
(703) 379-2480
www.agiweb.org

The American Geological Institute (AGI) is a nonprofit federation of 32 geoscientific and professional organizations, including the Soil Science Society of America. The AGI member societies represent more than 130,000 geologists, geophysicists, soil scientists, and other Earth and environmental scientists. Since its founding in 1948, AGI has worked with its members to facilitate intersociety affairs and to serve as a focused voice for shared concerns in the geoscience profession; to provide leadership for improving Earth-science education; and to increase public awareness and understanding of the vital role the geosciences play in society's use of resources and its interaction with the environment.

Soil Science Society of America
677 South Segoe Road
Madison, WI 53711
(608) 273-8095
www.soils.org

The Soil Science Society of America (SSSA) promotes basic and applied research in soils, fosters high standards in the teaching of soil science and the education of soil scientists, provides information on soil science to scientists and nonscientists alike, and promotes improvements in the field of soil science. The Society, established in 1936, has more than 5,700 members.

USDA, Natural Resources Conservation Service/
Soil Survey Division-National Soil Survey Center
100 Centennial Mall N, Room 152
Lincoln, NE 68508
(402) 437-5499
www.nssc.nrcs.usda.gov

The U.S. Department of Agriculture, Natural Resources Conservation Service (NRCS) works hand-in-hand with the American people to conserve natural resources on private lands. We help land users and communities approach conservation planning and implementation with an understanding of how natural resources relate to each other and to all of us and how our activities affect those resources.

Design and production: De Atley Design
Project Management: GeoWorks
Printing: CLB Printing Company

Copyright ©1999 by American Geological Institute.
All rights reserved.

ISBN 0-922152-50-0.

Contents

Foreword

Sustaining Our Soils and Society, the second publication in the American Geological Institute's Environmental Awareness series, was prepared under the sponsorship of the American Geological Institute's Environmental Geoscience Advisory Committee with support from the AGI Foundation. Since its appointment in 1993, the Committee has assisted AGI by identifying projects and activities that will help the Institute achieve the following goals:

✦ Increase public awareness and understanding of environmental issues and the controls of Earth systems on the environment;

✦ Communicate society's needs for managing resources, protection from Earth hazards, and evaluation of risks associated with human activities related to Earth processes and resources;

✦ Promote appropriate science in public policy through improved communication within and beyond the geoscience community related to environmental policy issues and proposed legislation;

✦ Increase dissemination of information related to environmental programs, projects, research, and professional activities in the geoscience community.

The objective of the Environmental Awareness series is to promote better understanding of the role of the Earth sciences in all aspects of environmental issues. Soil is a resource of critical importance and its use raises many environmental questions. We hope that *Sustaining Our Soils and Society* will help you identify and consider those questions.

Susan M. Landon
AGI President, 1998

Gary W. Petersen
SSSA President, 1998-99

Philip E. LaMoreaux
Chair, AGI Environmental Geoscience Advisory Committee, 1993-

Preface

Our objective in writing about soils is to relate them to environmental concerns that affect us all. Soil is a fragile, finite resource. It plays unique roles in maintaining air quality, storing water and nutrients for plants, and filtering contaminants from surface waters. Using soil as a medium for disposing of liquid and solid wastes raises environmental questions as does any soil-based construction project. Although this book discusses these topics, it focuses mainly on soil as society's primary source of food and fiber, such as cotton and wood. Growing plants in soil is the only known way to produce large quantities of these necessities and the implications of that fact are profound.

This book, like others in the American Geological Institute's Environmental Awareness series, discusses environmental concerns from a geoscience point-of-view. We, as citizens and policy makers, cannot afford to ignore the geoscience context, because Earth processes and systems affect all aspects of the environment.

In the United States, the number of people whose livelihood depends directly on the soil has decreased dramatically in the past 50 years. Those who live in urban settings, where opportunities for daily contact with natural environments and processes are limited, have few contacts with soil — or occasions to recognize its importance. Yet, life at the surface of the Earth evolved in soil, and the future of life depends on it. We hope that this booklet will increase your awareness and understanding of

- ✦ Why healthy soils are essential for our well-being
- ✦ The critical role soils play in a variety of natural processes
- ✦ The factors affecting the suitability of soils for various uses
- ✦ Management practices that conserve soils and keep them productive
- ✦ What the major sources of soil contamination are, and
- ✦ How damaged soils can be reclaimed

We are indebted to the many individuals who offered suggestions for improving the manuscript and for those who allowed use of photographs. We are especially indebted to Tom Hiett, Illustrator, Media Resource Center, Iowa State University, who prepared the original figures, and to Aaron Kitzman, Computer Specialist, Department of Agronomy, Iowa State University, who assisted in putting the figures on a World Wide Web page for review. Also, we gratefully acknowledge the assistance and valuable suggestions of Jerry Bingham, SSSA editor, Julie Jackson, AGI editor, and the outstanding graphic design by Julie De Atley, De Atley Design, Woodbridge, Virginia. Finally, we are indebted to the American Geological Institute Foundation, USDA, Natural Resources Conservation Serivce/ Soil Survey Division and the Soil Science Society of America for their support.

Thomas E. Loynachan
Kirk W. Brown
Terence H. Cooper
Murray H. Milford
November, 1998

It Helps to Know...

Desert soils (Aridisols) form in arid settings. Although this landscape is in Utah, the profile is from Arizona, where Desert soils are more widespread. Where irrigation water is not available, such soils are used mainly for range, wildlife, and recreation. Desert soils are commonly rich in calcium carbonate, which may form impermeable layers

Soil is a mixture of mineral particles, organic matter, chemicals, air, and water that supports life.

Dirt, ground, earth, soil — common names for the remarkably useful loose stuff that forms at the interface of Earth's rocky crust and the atmosphere. You can find soil almost anywhere on Earth except on steep, rugged mountains, areas of perpetual ice and snow, and in extreme deserts. Figure 29 shows the geographic distribution of soils in the United States. Where soil and moisture exist, life flourishes.

Because soil is so common, people tend to take it for granted — to forget that life as we know it could not exist without soil. Later chapters describe the many uses of soils, the ways some of them have been damaged, and methods of restoring, managing, and sustaining our soil resources. In this chapter, you'll find basic information about soils: what the environmental concerns are, what soil is, how and where it forms, and why soil quality matters.

What the Environmental Concerns Are

Our well-being — and the well-being of all who follow us — depends on society's effectiveness in managing soil resources. We can improve our chances of sustaining our soils and society by working locally and nationally to resolve the following environmental concerns:

- ✦ Maintaining agricultural productivity in light of world population growth
- ✦ Using nutrients to restore soil productivity without contaminating water supplies
- ✦ Addressing soil and use issues in their geoscience context
- ✦ Reducing erosion to conserve soils
- ✦ Maintaining quality of soils and water supplies
- ✦ Decreasing and preventing soil contamination, and
- ✦ Improving, restoring, and reclaiming damaged soils

Fig.1. *Profile of Nicollet loam soil west of Ames, Iowa*

What Soil Is

What is soil? The answer depends on who you're asking. In engineering and construction, soil is the usual name for earth that can be excavated without blasting. Geologists commonly use the term soil for a layer of weathered, unconsolidated material on top of bedrock (a general term for rock beneath soil). To soil scientists, soil is unconsolidated material at the surface of the Earth that serves as a natural medium for growing plants.

Soil is a mixture of mineral particles and organic matter of varying size and composition (Table 1). The particles make up about 50% of the soil's volume. Pores containing air and/or water occupy the remaining volume. A mature, fertile soil is the product of centuries of physical and chemical weathering of rock, combined with the addition and decay of plants and other organic matter.

The proportions of sand, silt, and clay in a soil determine its texture. In common usage the term "loam" (Fig.1.) refers to a medium-textured soil composed of sand, silt, and clay as well as organic matter. Loamy soils are usually well-drained and fertile; they are the best gardening soils. Top-soil, the soil at the surface, contains accumulated organic matter. Organic matter increases the fertility of topsoil, making it better suited for plant growth than the underlying subsoil. Subsoils tend to be less productive because they lack organic matter and commonly contain more clay.

Clay minerals and quartz, the two main products resulting from the weathering of rock, have important roles in soil development and plant growth. Quartz weathers to grains of sand and silt that help keep soil loose and aerated, allowing good water drainage. Clay minerals help to hold water and nutrients in a soil. Because most plants need air, water, and nutrients for optimum growth, the best soils

GRAIN-SIZE SCALE (Used by Soil Scientists)

mm	inch	Sieve Series		
2.0	0.08	No. 10		
			very coarse	Sand
1.0	0.04	No. 18		
			coarse	
0.500		No. 35		
			medium	
0.250		No. 60		
			fine	
0.100		No. 140		
			very fine	
0.050		No. 270		
				Silt
0.002				
				Clay

Table 1

Soil structure is the grouping of sand, silt, and clay particles into clusters. In topsoil, soil structure commonly resembles "bread crumbs."

for plant growth have a balance of clay, silt, and sand.

Air and water move through soil at rates determined by the size and arrangement of pores. Soils dominated by clay texture and blocky structure have more small pores and few large pores, so water moves slowly through them. The ideal distribution of large and small pores for plant growth is 50/50. The large pores (greater than 0.06 millimeters in diameter) allow water to infiltrate rapidly, and the small pores store water for plant use.

The organic matter in soils provides nutrients to plants and enhances the soil's ability to retain water. As organic matter decays (oxidizes) in the soil, nitrogen, phosphorus, and micronutrients are released. Soils high in organic matter (humus) are dark colored and usually form either under prairie grasses or in wet locations, such as bogs or marshes, where lack of oxygen slows decomposition of the organic matter.

Mineral soils with high concentrations of organic matter have excellent properties for plant growth. Excluding muck and peat soils, the amount of organic matter in mineral soils ranges from about 1 to 20% in the topsoil. Organic matter content decreases markedly with soil depth. Soils or soil layers having a very high content of organic matter (usually more than 20% carbon) are difficult to manage for agricultural and construction purposes. Such soils commonly are removed before construction begins, because organic matter compacts poorly and loses volume when it oxidizes.

Color in subsoils is determined, to some extent, by the availability of oxygen and its influence on forms of iron in the soil. Red and yellow hues indicate good aeration in the soil. When aeration is poor, soils turn gray or olive (Fig.2.). In some cases, only part of the soil is affected and the soil has splotches of both red and gray colors; this condition is called mottling. Mottled zones in soils tend to be seasonally wet, and little root activity takes place in these zones because oxygen for respiration is in short supply.

well aerated oxidized soil

mottled soil

poorly aerated compact soil

Fig. 2. *Air circulation affects color in subsoils.*

How Soils Form

Most soils take a long time to form. They form from rocks and sediments (parent material) that have disintegrated and decomposed through the action of weather and organisms (a process called "weathering"). The rate of soil formation is largely controlled by rainfall, temperature, and the type of parent material. High temperature and abundant rainfall speed up soil formation, but in most places a fully developed soil that can support plant growth takes hundreds or thousands of years to form.

Fig. 3. *Profile of a forest soil.*

Old soils are commonly more developed than young ones. Generally, old soils are less fertile because most of the nutrients have been leached (washed away). The oldest known soils are in Australia in a secluded rainforest canyon. The canyon is the home of the Wollemi pine, a tree whose origins date back more than 150 million years. In the United States, the oldest soils are only about one million years old; they lie upon high terraces of the Sierra Nevada mountains in California. In the northern one-third of the United States, soils are often less than 20,000 years old — the time when the last glacier retreated from the area.

The first stage in soil formation is the removal of soluble minerals and the accumlation of pieces of weathered rock and organic material. In the second stage, the soil particles are gradually rearranged as the soil matures, and distinct layers appear. The layers, called soil horizons, can be distinguished from one another by appearance and chemical composition (Fig. 3). Boundaries between soil horizons are usually gradual rather than sharp.

The top layer of soil, the A horizon, or zone of leaching, is characterized by the downward movement of water and the accumulation of decomposed organic matter. Water entering the ground moves downward, dissolving some of the soil materials and carrying them to deeper levels. In a humid (wet) climate, clay minerals, iron oxides, and dissolved calcite are most

Soil characteristics depend on

Source materials — physical and mineralogical composition of the parent material

Time — how long weathering has acted upon the parent material

Slope — soils tend to be thicker on flat land and thinner on steep slopes

Climate — soils tend to be thick in warm, wet climates and thin in dry ones

Biota — plants and animals in and on the soil, as well as human activity

typically moved downward. Leaching may make the A horizon pale and sandy, but the uppermost part is often darkened by humus (decomposed plant material) that collects at the top of the soil. This dark, upper layer is the topsoil.

The B horizon, or zone of accumulation, is a soil layer characterized by the accumulation of material moved downward from the A horizon above. This layer is commonly rich in clay and is colored red or yellow by iron oxides.

The C horizon is incompletely weathered parent material, lying below the B horizon. The parent material is commonly subjected to physical and chemical weathering from frost action, roots, plant acids, and other agents. In some cases, the C horizon is transitional between unweathered bedrock below and developing soil above.

The character of a soil changes with time. Over very long periods, the type of parent material becomes less and less important. Given enough time, soils forming in the same climate from many different kinds of materials can become quite similar.

With time, soils tend to become thicker; most modern soils have taken centuries to form. Another factor controlling soil thickness is the slope of the land surface. Soils tend to be thick on flat land, where more water moves into and through the soil, and thinner on steep slopes where water washes soil particles downhill.

Climate affects soil thickness and character. Soils in warm, wet climates, as in the southeastern United States, tend to be thick and are generally characterized by downward movement of water through the Earth materials. In arid (dry) climates, as in many parts of the western United States, soils tend to be thin and are characterized by little leaching, scant humus, and the upward movement of soil water beneath the land surface. The water is drawn up by surface evaporation and subsurface capillary action.

The evaporation of water beneath the land surface can cause the precipitation of salts within the soil. An extreme example of salt buildup can be found in desert saline soils, as in Death Valley, California, for example.

Measures of soil quality include how effectively soils:

- *accept, hold, and release nutrients*
- *accept, hold, and release water*
- *promote and sustain root growth*
- *maintain suitable biotic habitat*
- *respond to management and resist degradation*

Where Soils Form

Although soils may develop from the weathering of the rock directly beneath them, many soils are formed from parent materials transported from some other location. Such soils usually form in sediment deposited by running water, wind, or glacial ice. For example, mud deposited by a river during times of flooding can form an excellent agricultural soil next to the river after flood waters recede. The soil-forming mud was not weathered from the rock beneath its present location but was carried downstream from regions perhaps hundreds of miles away. Deposits transported by wind (loess) are the parent material for some of the most valuable food-producing soils in North America, Europe, and Asia.

Why Soil Quality Matters

Soil supports plant growth and represents the living reservoir that buffers the flow of water, nutrients, and energy through an ecosystem. Most of the water that people see and use falls first on the land. It then either percolates to the groundwater, runs over the land surface to a stream or lake, or moves laterally through the soil to a surface water body. No matter which route the water takes, the quality of the soil largely determines the chemical and biological characteristics and flow dynamics of the water passing through it.

The capacity of a soil to support plant growth and act as a buffer is a measure of its quality. Soil texture, structure, water-holding capacity, porosity, organic matter content, and depth are some of the properties that determine soil quality. A soil with sufficient capacity to support one ecosystem — rangeland, for example — may not be capable of supporting another, such as a corn field.

Soil quality is important for two reasons. First, we should match our use and management of land to soil capability, because improper use of a soil can damage it and the ecosystem. Second, we need to establish a baseline understanding about soil quality so that we can recognize changes as they develop. By using baselines to determine if soil quality is deteriorating, stable, or improving, we have a good indicator of the health of an ecosystem.

Why Soils are Important

Organic soils (Histosols) are wetland soils, dark in color and rich in decomposed organic material. They form in poorly drained and lowland environments in the Great Lakes region and coastal eastern United States. Field crops growing in Michigan illustrate a fertile organic soil. The organic soils in wetland areas play an important role in environmental protection by filtering contaminants from surface water — especially during periods of high surface runoff and flooding

Food and Fiber

Without soils neither we nor the ecosystems in which we live could exist. Even the survival of fish and other aquatic organisms depend on nutrients released from soil into streams and oceans. Most food and fiber come from the soil. To give some idea of the scope of agriculture in the United States, Americans produce at least two hundred different crops. Hay, wheat, corn, and soybeans account for about 80% of the acreage planted each year, and beef and dairy cattle, poultry, and hogs, which eat these crops, account for slightly more than half of the total value of all farm sales. Major fiber products include timber, cotton, wool, and hides.

Fig. 4. *The many roles of soil*

Soils have many uses in addition to food, fiber, and fuel production. They play a major role in recycling carbon to the atmosphere and nitrogen in the soil, storing water for plant use, filtering surface waters, and in disposal of solid and liquid wastes (Fig. 4). Soils are also the base beneath most of our homes and roads as well as an important source of building materials, such as adobe.

Carbon Recycler

The chemical element carbon (C) is a major component of all living things. The chemical, physical, and biological properties of soils make them an ideal medium to support the growth of plants as well as the decay of dead plants and animals. One of the main functions of microorganisms in soil is to oxidize carbon to carbon dioxide (CO_2). Oxygen (O_2) in the soil is used rapidly by roots and microbes that give off CO_2 during the decomposition process. Although CO_2 levels in the atmosphere are only about 0.03%, this small fraction is critical to life. An estimated 85% of the atmospheric CO_2 comes from biological oxidation reactions in soil. As dead plants and animals decompose they complete the carbon cycle (Fig. 5) by releasing CO_2 into the atmosphere.

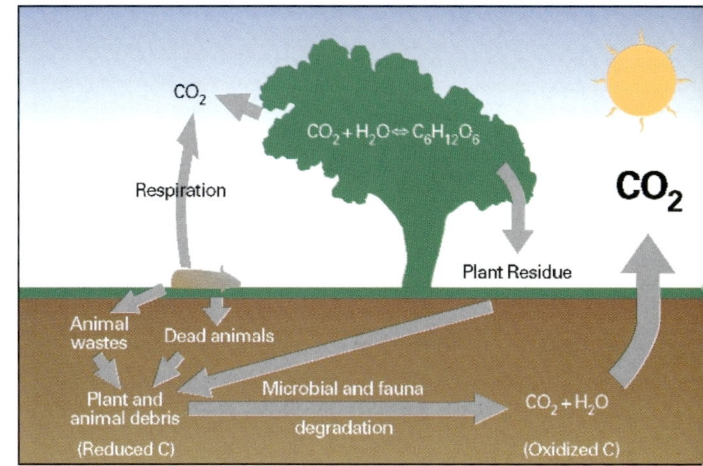

Fig. 5. *The carbon cycle centers around atmospheric carbon dioxide (CO_2), which is processed and reduced by organisms using energy from sunlight; oxidized by plants, animals, and microorganisms in the soil; and released into the atmosphere.*

Organically bound carbon, in the form of plant residues and animal wastes and bodies, eventually falls to the soil and breaks down or decomposes. A small amount of the resistant carbon remains in soil as humus, the dark

Carbon dioxide is considered a "greenhouse gas" and increased levels of CO_2 and other greenhouse gases in the atmosphere may contribute to global warming. The major greenhouse gases and their overall contributions to the atmosphere are calculated as

carbon dioxide (CO_2)	*60 %*
methane (CH_4)	*15 %*
nitrous oxide (N_2O)	*05 %*
ozone (O_3)	*08 %*
chlorofluorocarbons	*12 %*

(Pierzynski et al., 1994)

organic part of the soil. Estimates indicate that carbon held in soil humus is about twice that present in the atmosphere as CO_2, and that Earth's soil organic reservoir stores as much as three times more carbon than all of the planet's vegetation. Other carbon gradually converts into forms that are stable, such as carbon in peat bogs, coal, and oil deposits. Still other carbon is in long-living vegetation such as mature trees; that carbon is slowly recycled back to the atmosphere as trees die and gradually decompose into soils.

Plants absorb carbon dioxide during growth, and they are an important repository for it. They use carbon dioxide, water, and chlorophyll to photosynthesize or transform radiant energy from the sun into sugars and other chemicals needed for growth. A rising level of atmospheric carbon dioxide is expected to increase the photosynthetic productivity of soils. In fact, commercial greenhouses commonly add carbon dioxide to the air to increase yields of cucumbers, tomatoes, and some ornamental plants.

Removal of Greenhouse Gases

Greenhouse gases, such as carbon dioxide and methane, act as an insulating blanket around the Earth, allowing incoming solar radiation to warm the Earth's surface and reducing radiation of heat back through the atmosphere into space (Fig. 6). The blanket of soil covering the Earth's surface is thought to be important in mediating the effects of greenhouse gases. Furthermore, soil bacteria are known to oxidize methane (CH_4) and nitrous oxide (N_2O) to less harmful end-products. On the other hand, wet soils and swamps are important natural sources of methane.

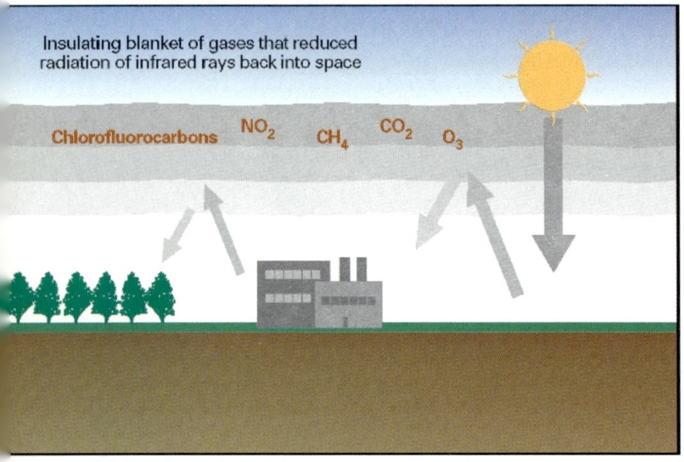

Fig. 6. *Increased levels of "greenhouse gases" in the atmosphere may contribute to global warming.*

Nutrient Recycler

Plants require 16 essential nutrients for growth. Three of these nutrients, carbon, hydrogen, and oxygen, come from carbon dioxide and water. The remainder come from the soil. The major fertilizer elements are nitrogen, phosphorus, and potassium. Plants also require large quantities of calcium, magnesium, and sulfur as well as smaller quantities of iron and other micronutrients. As minerals weather, they release some of these nutrients, but many nutrients come from the breakdown and recycling of

Thirteen of the 16 nutrients which growing plants require originate from soil. Growing plants need large amouts of nitrogen, phosphorus, potassium, calcium, magnesium, and sulfur. They require smaller amounts of iron, copper, manganese, boron, chlorine, zinc, and molybdenum for growth.

residues. By recycling the nutrients contained in leaf litter, dead roots, and animal wastes, a natural ecosystem helps maintain new growth.

In agricultural fields, some of the products of each season are harvested and removed from the field, rather than left to decay and return their nutrients to the soil. If left without proper replenishment, a cultivated soil tends to use up its initial reserves and to be depleted of readily available nutrients after a number of years, so a gradual loss of fertility occurs. Fertility can be maintained, however, by the regular addition of organic manures or balanced mineral fertilizers. Modern agriculture relies on continuous applications of mineral or organic fertilizers to maintain yields.

Water Storage

Soils store water and make it available to plants for use during dry periods. In most areas of the world, the water needs of plants during the main growth period exceed the amount of rainfall. Plants use tremendous amounts of water while growing — 180 to 900 kilograms (400 to 2000 pounds) of water for every kilogram (2.2 pounds) of dry matter produced. The water requirement varies with soil conditions and climate. In a fertile soil in a moist climate where growing conditions are good, the amount of water needed to produce a kilogram of dry matter is much less than in a poor soil with inadequate moisture supply. Approximately 40 centimeters (16 inches) of water is needed to produce an average corn crop in the Midwestern United States (Fig. 7).

A soil must have a large water-holding capacity to store enough water for plant growth. The soils best-suited for this purpose are well aggregated (made up of particles that stick together as a unit) and have a medium texture and adequate levels of organic matter. Such soils absorb precipitation rapidly and allow it to move internally. Where precipitation is inadequate or the soils are sandy and have low water-holding capacities, irrigation may be needed to grow plants.

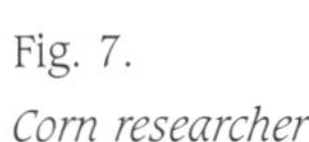

Fig. 7.
Corn researcher

Steps in the water cycle

- ✦ ***Precipitation*** — water falling on the Earth from rain or snow
- ✦ ***Overland flow*** — water moving off the land via streams or rivers
- ✦ ***Infiltration*** — water entering the soil and deeper strata (recharge)
- ✦ ***Evaporation*** — water lost from the land surface
- ✦ ***Transpiration*** — water lost from plants through their leaves
- ✦ ***Groundwater storage and discharge*** — water stored in the Earth that discharges into springs, streams, lakes, and oceans

During precipitation, plants and soils absorb a portion of the water. Some of the water flows into streams and some evaporates. As the water soaks deeper and deeper into the ground, it infiltrates and recharges aquifers, the underground layers of porous rock and sand that store water. Groundwater moves through the Earth as part of a dynamic flow system from recharge areas to discharge areas such as springs. Soil characteristics as well as the duration and intensity of precipitation determine the rate at which water enters the ground and reaches the water table (the level below which the ground is saturated with water) or runs off the surface into wetlands, lakes, rivers, or oceans. Evaporation moves water back to the atmosphere. Precipitation, overland flow, infiltration, evaporation, transpiration, and groundwater storage and discharge are steps in a continuous, dynamic process called the water cycle, or hydrologic cycle (Fig. 8).

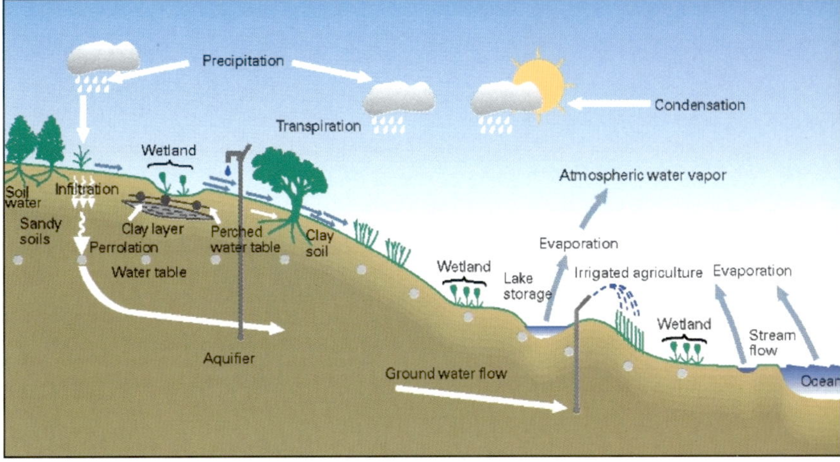

Fig. 8. *The water cycle and the role of the soil. Sandy soils will have more infiltration and less runoff than clay soils. Perched water tables will create wetlands above the natural groundwater aquifer.*

Water Filter

Wetlands, a general term for marshes, swamps, and other areas that are permanently wet and/or intermittently water-covered, are particularly good water filters (Fig. 9). Typically, wetlands occur at the mouths of river valleys, along low-lying coasts, and in valleys. Some occur even on high plateaus.

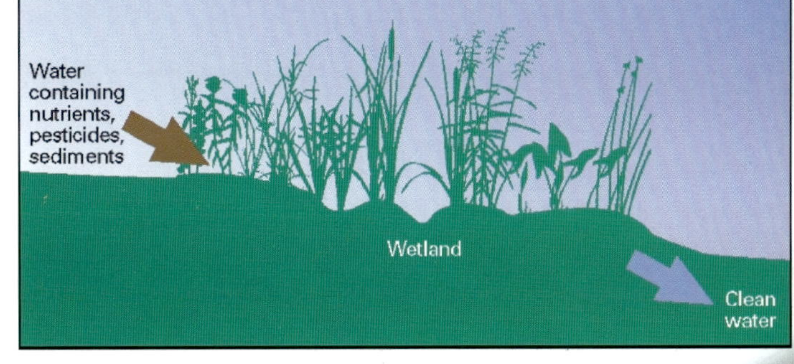

Fig. 9. *Wetlands act as filters to clean surface waters.*

Wetlands help maintain and improve the water quality of rivers and other water bodies by removing and retaining nutrients, processing chemical and organic wastes, and reducing sediment loads to receiving waters. When nutrients and contaminants are removed from upland areas by soil erosion or groundwater flow, wetlands may intercept the surface-water runoff and filter it. The filtering capacity of wetlands is especially important for removing nitrogen and phosphorus. Where filters such as wetlands are not present, these nutrients move directly into lakes and rivers, causing excessive growth of aquatic plants and algae. As a result, water in the lake or pond may become depleted enough in oxygen to kill fish and other aquatic life.

Wetlands also hold flood waters that overflow riverbanks as well as surface waters that collect in isolated depressions. Wetland vegetation helps to slow the speed of flood waters, keeping more water out of the rivers. Wetlands within and upstream from urban areas are especially valuable for flood protection, since urban development increases the rate and volume of surface-water runoff, thereby increasing the risk of flood damage.

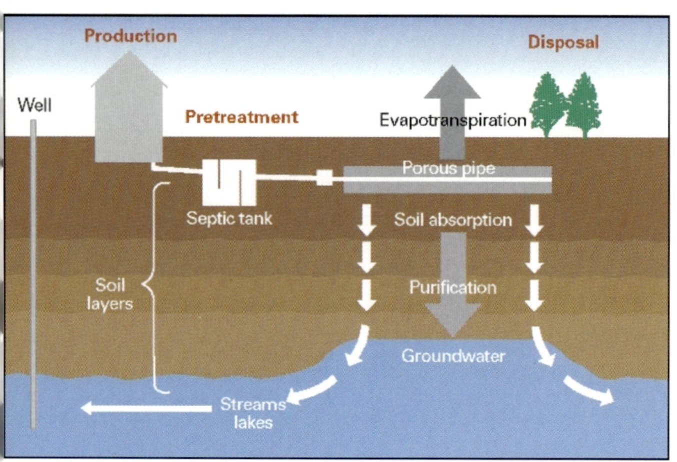

Fig. 10. *Soil is used to filter wastewater coming from rural homes before reaching groundwater or drinking wells (Redrawn from Bouma et al., 1972).*

Waste Disposal

Disposal of wastes by mixing them into surface soil is called "land treatment." Sewage sludge; food-processing sludge; yard waste such as leaves, lawn clippings, and branches; composted garbage from which glass, metal, and plastic have been removed; oil; and oil-contaminated soils are good candidates for land treatment. Other waste materials include waste gypsum from wallboard, lime kiln dust, and flue-gas desulfurization by-products. Controlled amounts of each of these wastes can be safely and effectively mixed with soil, resulting in the recycling of nutrients that the wastes contain.

The U.S. Environmental Protection Agency (EPA) has developed programs and set standards to encourage land treatment of properly processed sewage sludge. The processing involves destruction of disease-causing organisms and stabilization to minimize odor problems. According to EPA figures, one-third of the treated sewage sludge from publicly owned treatment works in the United States is presently being applied to soils. Annual application rates of sewage sludge are selected to supply plant nutrient needs.

Homes not connected to municipal sewage treatment facilities must treat their wastes on site. The average person in the United States uses about 200 liters (50 gallons) of water per day for washing, bathing, cooking, and sanitary purposes. Commonly, the wastewater passes into a septic tank and then into a soil filter or leaching field (Fig. 10). The suitability of a soil to purify the wastewater in the leaching field is determined by several factors, including flooding potential; depth to water table, bedrock, or restrictive layers; drainage; permeability; texture; slope; and soil slippage. Any of these features can cause the soil to be unsuitable. A major problem in some locations, especially around lakes bordered by houses, is septic systems that partially or totally fail. When this happens, the untreated wastewater passes into the lake, causing algal blooms and eventual lake degradation.

Soil also is used as a component of the liners and final covers for landfills. The complex subject of landfills is covered by extensive regulations to handle a variety of geologic settings, soil conditions, and environmental questions. The environmental objective is to minimize the leakage of clean water into the landfill and the leakage of contaminated water, called leachate, from the landfill into the groundwater. Most soils are porous and permeable, so only soils with high clay contents, which can be compacted to reduce their permeability and, more importantly, the size of the pores in the soil, can be used to line or cover landfills (Fig. 11).

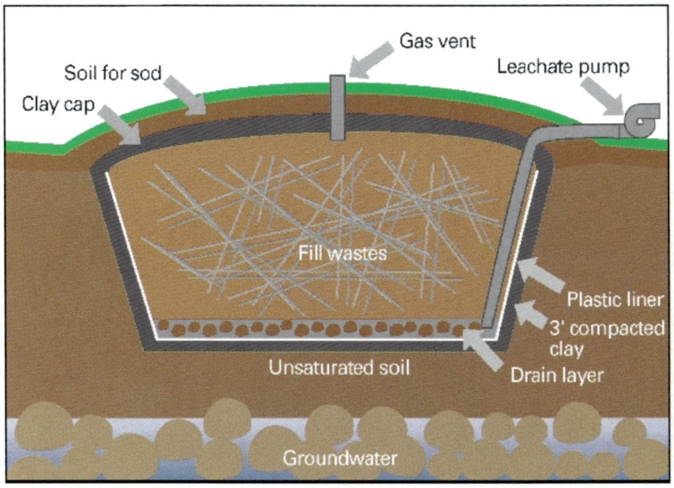

Fig. 11. *Generic design of a modern landfill. To limit water movement into and through the waste materials the selected soil is excavated, broken up, brought to an optimum water content, and compacted in thin layers. The compacted layers are built up, one on top of another until a 0.5-1.0 m thick (2 or 3 ft) soil liner is achieved. After the landfill is full, a similar system of compacted soil layers is placed over the waste.*

Use in Construction

Most buildings, roads, dams, and pipelines are built on soils. A well-drained, sandy or gravelly soil provides the most trouble-free base for roads and buildings. If soils are not well drained, drain tiles are laid or the land surface is graded to move water away from buildings. Side ditches are used to move water away from roadbeds (Fig. 12).

Some building materials, for example, adobe and bricks, are made from soil. The low thermal conductivity of soil makes it an ideal insulating material for energy-efficient buildings that stay cool in summer and warm in winter. The early settlers on the Great Plains of the United States, where wood was scarce, commonly built their homes from sod. Adobe is a common building material in warm, dry climates; it is made from a mixture of straw and moist soil that becomes hard when dry. Houses of mud brick and rammed earth are other examples of the usefulness of soil as a building material. Today, new construction in many climates takes advantage of the insulating properties of soil by putting buildings partially or fully underground.

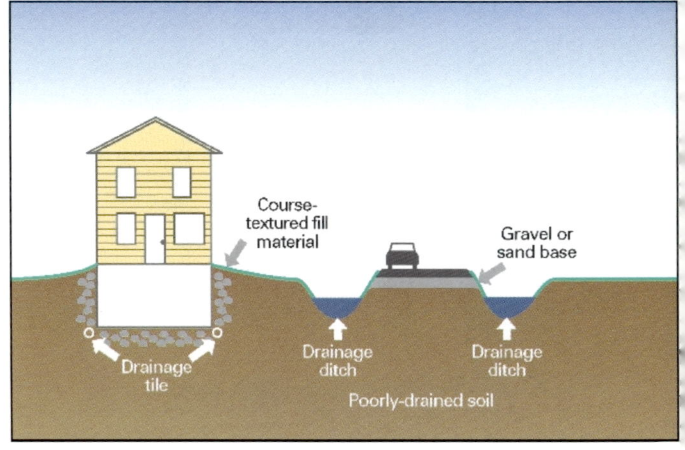

Fig. 12. *Drainage is important to remove excess water through coarse-textured materials.*

The physical and chemical properties of soils are critically important for construction and engineering applications. The next chapter discusses those properties and compares the suitability of soils for a variety of purposes.

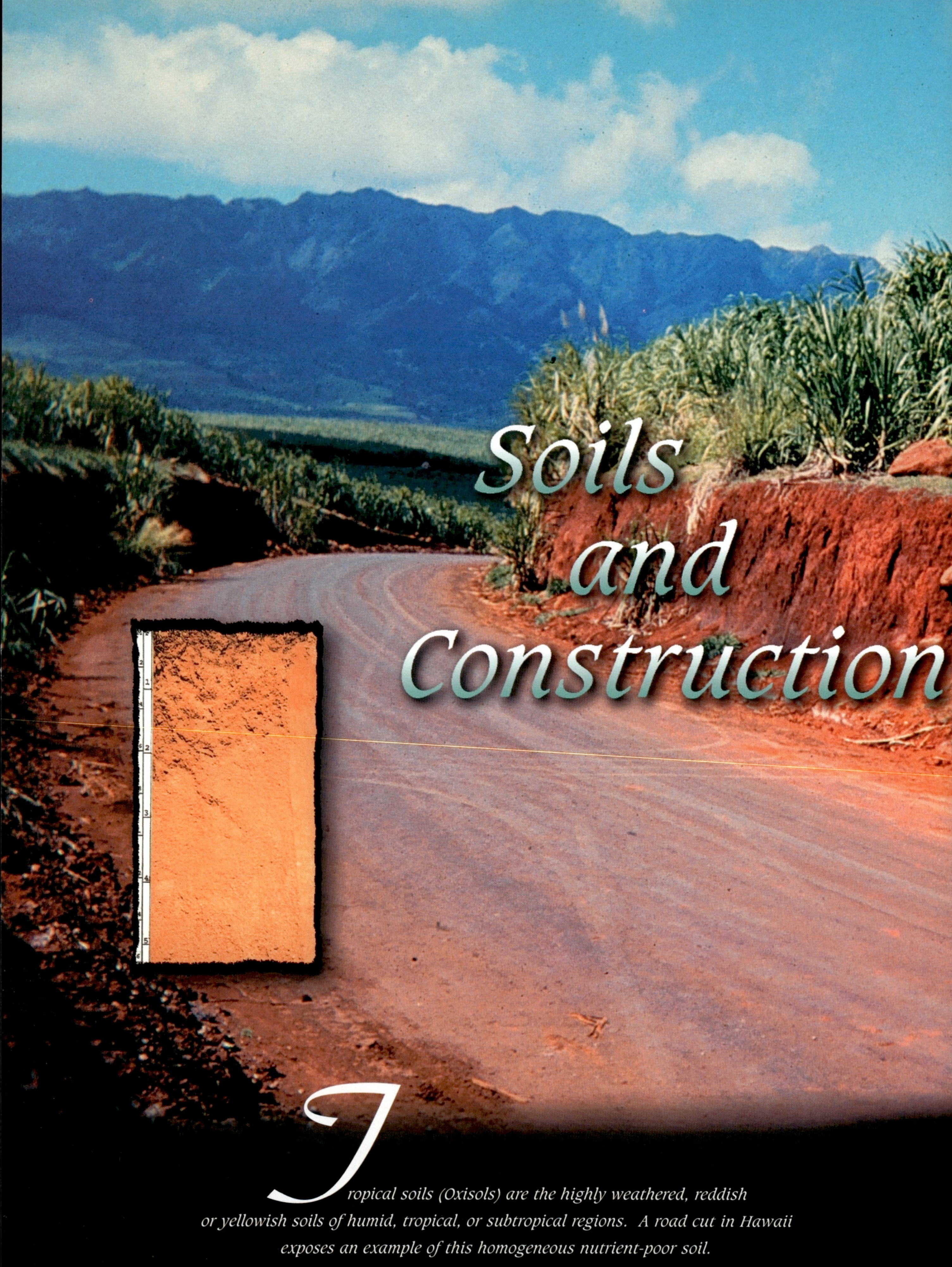

Soils and Construction

Tropical soils (Oxisols) are the highly weathered, reddish or yellowish soils of humid, tropical, or subtropical regions. A road cut in Hawaii exposes an example of this homogeneous nutrient-poor soil.

Matching soils to the engineering and construction uses for which they are best suited protects the environment — and saves money.

The success of any soil-based construction project depends on the suitability of the soil for the intended use. Analysis of the soil's physical, chemical, and engineering properties before construction begins is critical, because soil characteristics commonly vary across short distances and with depth. Engineering properties of soil, such as texture, structure, and strength, are especially important in construction applications. The size and relative abundance of sand, silt, and clay particles in a soil determine whether its texture is coarse, medium, or fine. Medium-textured soils have the most desirable engineering properties, and, in general, extremely coarse or fine soils are less suitable.

ENGINEERING AND CONSTRUCTION PROPERTIES

Coarse sandy soils

- *Don't hold sufficient water or nutrients for plant landscaping*
- *Are less susceptible to erosion because water infiltrates rapidly*
- *May need special footings for foundations to distribute weight*
- *Provide little filtration for septic wastewater disposal*
- *Cannot be compressed and compacted*

Fine-textured clay soils

- *Bind water and nutrients too tightly for plant landscaping*
- *Are highly susceptible to erosion by water because water infiltrates slowly, in soils with poor structure*
- *Provide poor foundations if the soil contains expanding clay*
- *Provide low infiltration for septic wastewater disposal*
- *Produce poor roadbeds if the soil contains expanding clay*

Besides texture, two other properties greatly affect soil use: structure and type of clay. Structure is the grouping together and binding of sand, silt, and clay particles into larger aggregates (Fig. 13). A fine-textured soil that has good structure may be as porous as a coarse-textured soil because of the large pores present among aggregates.

Fig. 13. *Upper photograph shows a surface soil with good structure. Lower photograph shows a soil where falling rain drops have destroyed soil structure and caused the soil to form a "crust" (T. Loynachan).*

Fig. 14. *A soil high in smectite clay, such as this one in California, swells when it becomes wet and shrinks and develops large cracks as it drys. The notebook is approximately 9 by 14 cm (3.5 by 5.5 inches) (W. Lynn, NRCS).*

Clays

Clay is a universal constituent of soils, and the type of clay can have significant impact on land use. For example, expanding clays (known as smectites, Fig. 14) which swell as they absorb water are very desirable for sealing lagoons or dams, because they remain swollen when wet and do not allow much water seepage. However, because expanding clays shrink and crack as they dry, they make unstable foundations and roadbeds and cause major damage to them.

A soil survey and laboratory analysis can identify expanding clays. Soils with a high content of shrink-swell clays may need to be excavated before construction begins. Foundations can be ruined in just a few years if this precaution is not taken (Fig. 15). Besides cracked foundations, the effects of shrink-swell clays are seen as cracked highways and streets, cracked walls and window frames, doors that do not close, displaced and broken gas and water pipes, and leaning telephone and electric poles. Little can be done to improve soils with a high content of shrink-swell clays. If they cannot be avoided, precautions must be taken to design stronger structures or use special engineering designs to minimize their effects. Homes built on expanding-clay soils generally require foundations of re-enforced concrete.

Fig. 15. *Cracks in brick wall resulting from the failure of a foundation, caused by soil with a high content of expanding clays (D. Williams, NRCS, retired).*

TESTS FOR SHRINK-SWELL CLAYS

Simple tests can help detect the presence of clays that swell as they absorb moisture and shrink as they dry. A soil survey and laboratory tests can also identify them.

✦ As the soil in the field dries, do cracks readily form? Some surface soils high in shrink-swell clays may have cracks at the surface that are 10 cm (4 inches) across or greater.

✦ At a construction site, if a soil horizon high in shrink-swell clay is suspected, take a sample of the soil, moisten it, and make a molded sphere (perhaps 8-10 cm across). If the sphere fractures and cracks as it dries, the soil probably is high in shrink-swell clay.

Soil Strength

Construction engineers must consider the strength of a soil, which is its resistance to deformation or collapse. Soil strength relates directly to the ability of a soil to support a load. Water holds soil particles together, and they resist movement by frictional resistance over rough, uneven surfaces. Thus, soils with varying textures have different strengths at different soil water contents.

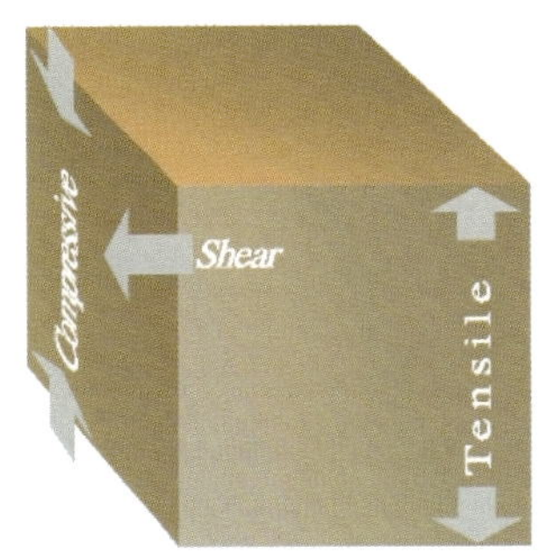

Fig.16. *Soil tensile, compressive, and sliding (shear) strengths.*

Soil strength (Fig. 16) can be assessed by the forces (stresses) required to pull the soil apart (tensile strength), push the soil together or compact it (compressive strength), or shear the soil (sliding strength).

In unsaturated soil, compaction is the most obvious and simple way of increasing the stability and supporting capacity. With the loss of pore space, the voids become smaller, and as particles have increased surface contact, the bonding between particles increases.

Each soil has an ideal water content for maximum compaction. Soils that are too wet, or too dry, do not compact well. Water reduces the compaction potential of soil. Because a high water content can serve as an internal lubricant, the same soil that is difficult to compact may be easy to shear (low shear strength). A dry soil, on the other hand, has relatively high shear strength due to strong bonding and high internal friction, and thus cannot be readily compacted.

Engineers use a classification system based on mechanical properties of soils at different water contents. A dry soil will shatter when struck with a hammer, whereas a moist soil may be more plastic and only deform. As the soil becomes wetter, it may no longer hold its shape when molded. The water content of soils at these limits helps an engineer determine how suitable a soil is for various uses.

A beach is a good place to observe the influence of water on compressive strength of sand. While standing in the water, your toes sink easily into the sand. It doesn't support your weight completely. When you leave the water, there is a point where the sand readily supports your weight and you leave few tracks. As the sand gets drier, your toes again sink into it.

Food and Fiber Production

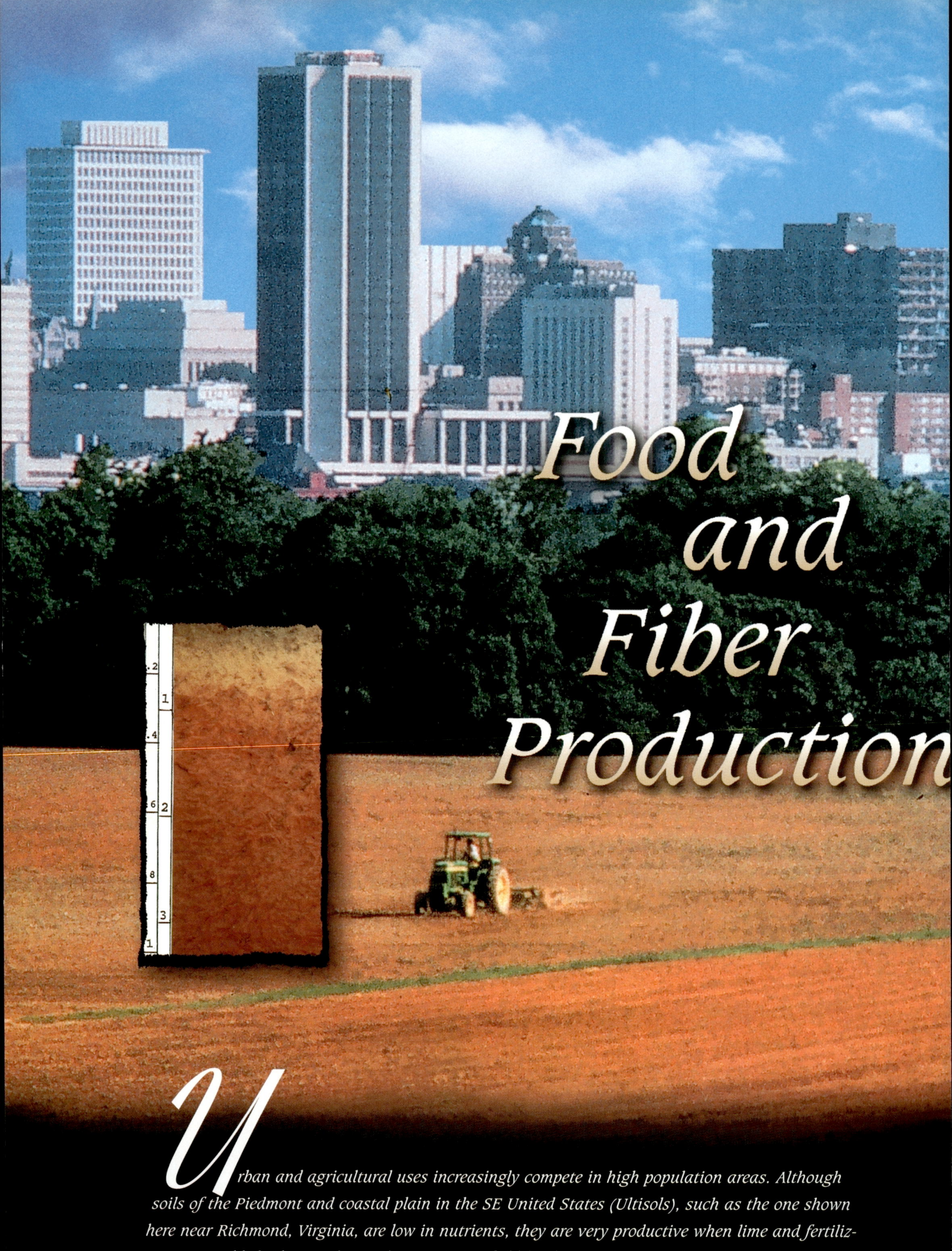

Urban and agricultural uses increasingly compete in high population areas. Although soils of the Piedmont and coastal plain in the SE United States (Ultisols), such as the one shown here near Richmond, Virginia, are low in nutrients, they are very productive when lime and fertilizers are added. These soils are characteristic of old landscapes in warm, humid climatic zones.

World population has more than tripled since 1900, rising from 1.6 billion to 5.9 billion. The U.S. Census Bureau predicts that world population will reach 6.1 billion by the year 2000.

If you think of soil as a biological and chemical factory that supplies the food and fiber we need to survive, it is easy to see why good management is essential. Good management promotes healthy soils in which life thrives. Poor management leads to damaged soils in which life barely survives. Properly managed soils can support vegetation, produce crops year after year, and provide clean water for streams and lakes. Some soils in Europe, in Denmark and France, for example, have been in continuous cultivation for centuries with little evidence of reduction in their ability to yield crops or to protect the environment.

Early humans relied on natural fruits, vegetables, and animals that the soil produced, harvesting what they needed. As populations and demands for food increased, our ancestors developed ways to cultivate and manage soils. Their techniques of soil management — cultivation, fertilization, irrigation, drainage, and erosion control — form the basis for modern industrial agriculture. This industry produces most of the food, fuel, and fiber needed to sustain human life. Although dramatic increases in soil productivity have been achieved, increases in world population have kept pace with, or exceeded, the increases in productivity.

Population and Land Use Trends

World population has more than doubled since 1950, and it is growing at the rate of approximately 90 million people per year. The U.S. Census Bureau predicts that world population will reach 6.1 billion by the year 2000. Increasingly, people are moving to urban areas. Many urban areas are expanding and building on land that was once prime farmland. As as result, total cropland in the United States is decreasing (Fig. 17). Nearly 11% of the land in the United States has been converted to nonagricultural uses. The U. S. Department of Agriculture Statistical Reporting Service estimates that on a worldwide basis, 20 to 30 million hectares (49.4 to 74.1 million acres) will be converted to urban uses annually between 1980 and 2030.

Fig. 17. *Decline in total cropland in the United States from 1982 through 1997 (not drawn to proportion)*
(Natural Resources Inventory - 1997 State of the Land Update, USDA-NRCS)

Less than 12% of the world's land area is used to grow and harvest crops, and the amount of cropland with highly productive soils is shrinking. About 20% of the land in the United States is cropland. Since the 1920s, changes in agricultural markets, technology, and practice have dramatically affected the location and use of this land. Millions of acres of former cropland now support forests in the northeastern and southern states. And, crops are now grown on what was once Mississippi River bottomland forest and Great Plains grassland. Millions of other cropland acres have been converted for residential, business, and industrial uses.

Most of the world's potential new cropland is in South America and Africa. Little potential new cropland remains in Southeast Asia and Europe. Because of expanding world populations and the increasing use of land for nonagricultural purposes, the amount of grainland per person has declined from more than half an acre in 1950 to less than one-third acre in 1990. Brown and Kane (1994) predict that by 2030 the amount of grainland per person will be only 0.08 hectare (0.20 acre). It is already approaching this value in the more densely populated areas of the world. China, for example, has 0.1 hectare (0.25 acre) of grainland per person and little opportunity to increase cropland. As the amount of cropland declines, we must find ways to make and keep soils highly productive.

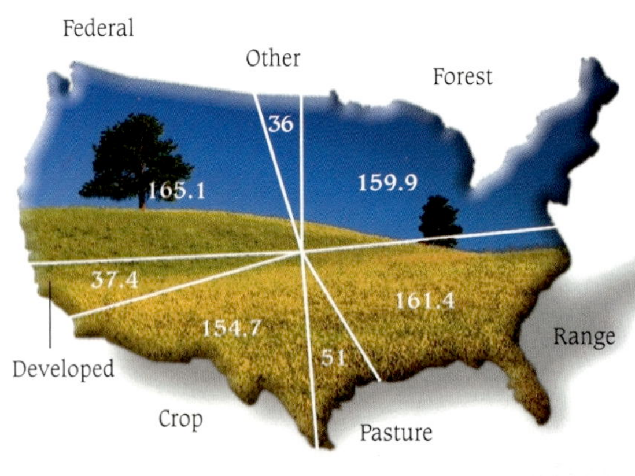

Fig. 18. *U.S. land use, 1992, in millions of hectares*
(USDA Soil Conservation Service, 1994)

Changes in land use (Fig. 18) obviously affect the soil, the landscape, and the environment. Urban streets and rooftops change drainage patterns so that storm water runs directly into sewers and drainage ways and is no longer filtered naturally through the soil. Urbanization brings new sources of pollution and contamination, such as oil leaked from automobiles and chemicals leached from suburban lawns. Urbanization also affects watersheds, the areas drained by rivers or river systems, and makes the task of developing effective, cooperative efforts to maintain healthy conditions in watersheds all the more difficult and necessary.

The first step in helping to ensure that land-use changes are not harmful is to evaluate current land-use trends and assess how well the basic natural resources — soil, water, air, plants, and animals — are faring. Good evaluation and assessment enable landowners to use and manage their land within its capabilities.

THE ENVIRONMENTAL CHALLENGE

World population is increasing at the rate of 90 million people per year, and the amount of agricultural land is decreasing. The challenge is to increase soil productivity with minimal damage to the environment.

A natural ecosystem is a community of animals, plants, and microorganisms, with interrelated physical and chemical environments.

Soil, the Plant Growth Medium

Life thrives in the biosphere, the zone at the interface of Earth's rocky crust and the atmosphere. Here, the sun provides radiant energy, which green plants capture and transform into sugars and other useful chemicals through the process of photosynthesis. Moist soils provide nutrients, water, oxygen, and physical support to plants — everything plants need except the energy from the sun and carbon dioxide from the air. As organisms in the soil decompose dead plants and animals, carbon dioxide is returned to the air, completing the carbon cycle (Fig. 5).

In a natural ecosystem, a community of animals, plants, and microorganisms with interrelated physical and chemical environments, residues from organisms are returned to soils. The soils in turn supply living organisms with nutrients. Such cycling of nutrients is a requirement of all ecosystems, and the organisms in soils play a crucial role in decomposing residues and converting nutrients from organic to inorganic forms. As the cycle of life continues, inorganic forms of nutrients can be returned to organic forms in the tissues of the next generation of plants and animals. However, in urban settings and other densely populated areas, the waste products from plant and animal products rarely are returned to their soil source areas. The waste products commonly are destroyed at treatment plants, buried in landfills, or added in concentrated amounts to land areas near animal or human population centers — sometimes with adverse effects on the environment.

In managed systems, the challenge is to maintain a nutrient balance. The goal is to provide an adequate supply of nutrients to plants and soil organisms and at the same time minimize the loss of nutrients from soils by erosion, evaporation, surface runoff, or leaching (the loss of dissolved matter in water). Application of animal manures, sewage sludge, and other organic wastes to land is practiced widely, and these practices can be very beneficial to soil productivity. However, care must be taken to avoid contamination from possible heavy metals, toxins, or salts in the waste materials. The environmental consequences of excess nutrients may be pollution of surface waters and pollution of groundwater. Ideally, the nutrient supply in soils should balance the nutrient demands of the plants, animals, and microorganisms in the system.

Productivity

The ability of a soil to support plant communities under specified systems of management is a function of many factors including the water-air balance in pores within the soil, nutrient supply, microbial activity, and rooting volume. Highly productive soils for most crops, such as corn and wheat, are those that are level to very gently rolling with an adequate supply of nutrients, water, and air in the root zone (Fig.19). Rice is an exception for it can grow on flooded soils that have poor aeration.

Fig. 19. *Strip farming near London, Ontario. Crops growing in strips are maize (corn), wheat, and soybeans. Alfalfa is grown in a strip across the ends of the other strips.*

Farmers in much of the developing world have sustained soil productivity for centuries by practicing forms of shifting cultivation (Fig. 20). They plant crops in new land and give the old fields time to rest and rejuvenate with the return of native vegetation. The prospect of producing enough food for growing populations from these traditional systems is bleak. Population pressures force farmers to return to idle land sooner, giving it less time to rejuvenate. When insufficient biomass (the total mass or amount of living organisms in a particular area or volume) accumulates above ground to provide adequate nutrients, fertility decreases. The lowered productivity puts even greater stress on the system (Brady and Weil, 1999). It is estimated that shifting systems of farming fill the food and fiber needs of about 5% of the world's people.

Fig. 20.

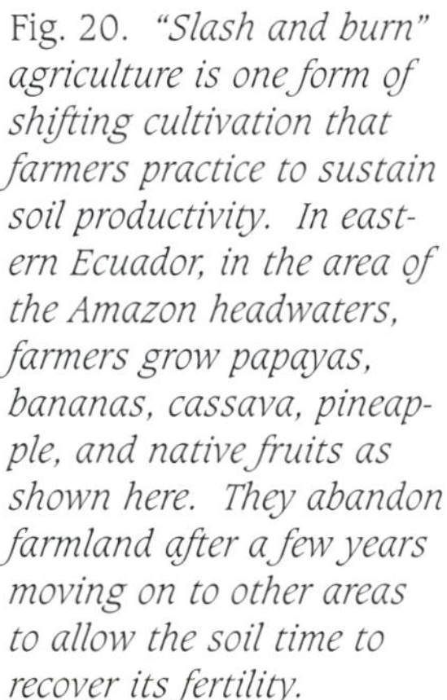

Fig. 20. *"Slash and burn" agriculture is one form of shifting cultivation that farmers practice to sustain soil productivity. In eastern Ecuador, in the area of the Amazon headwaters, farmers grow papayas, bananas, cassava, pineapple, and native fruits as shown here. They abandon farmland after a few years moving on to other areas to allow the soil time to recover its fertility.*

Precision Farming

Precision farming, or site-specific management, is an approach for balancing nutrient supply and demand, and for adjusting other major management inputs, such as pesticides, water, and tillage to varying soil conditions. For example, it may be more effective to add smaller, carefully timed applications of fertilizer than to put it all on a field at once. Taking advantage of technology, a farmer can get computer-generated maps of the landscape from natural resource databases generated by geographic information systems (GIS). In addition, tractors and other moving machines can be equipped with instruments that use data from satellites, the global positioning system (GPS), to provide continuous accurate geographic location (Fig.21). Using GIS and GPS in combination, a farmer can determine the precise amount of fertilizer, pesticides, seed, water, and other agricultural inputs needed and apply them for each set of diverse geographic conditions across a field using highly technical, computer-controlled machinery.

Fig. 21. *Computer-controlled machinery equipped with instruments that use a satellite guidance system (GPS) help farmers determine precise amounts of fertilizer, pesticides, and seed to apply to maximize productivity and minimize environmental impact.*

Additional steps for maintaining productivity include using herbicides and crop rotations to control weeds and other pests, using specially developed equipment for sowing seeds into surface crop residues, and planting without tilling (Fig. 22) or with minimal tilling. These practices protect soil surfaces from erosion and help sustain a higher organic content in the soil. They can help support global needs for food and fiber by maintaining soil productivity over long time periods with minimal adverse environmental impact.

Fig.22. *Maize (corn) planted in a fescue sod in Kentucky.*

Today, each acre of cropland produces nearly 3 times what was produced on the same acre in 1935. But farmers and ranchers produce much more than food and fiber.... Well-managed agricultural land also produces healthy soil, clean air and water, wildlife habitat, and pleasing landscapes, all of which are increasingly valued by rural and urban citizens alike.

— *America's Private Land, A Geography of Hope*

Environmental Consequences of Mismanagement

Excessive erosion by wind and water reduces soil productivity and contaminates air and water.

The consequences of soil mismanagement are severe and sometimes irreversible. Erosion is a natural process that wears away the land, but mismanagement leads to excessive erosion and other kinds of soil degradation that reduce soil quality and productivity. As soils become less productive, it becomes increasingly difficult and costly to sustain them and the living organisms they support. Erosion, salinization, the depletion of organic matter and nutrients in agricultural soils, and soil contamination all result from mismanagement. In some cases, mismanagement has reduced the growth of plants to the point where the yield of seed is less than that planted.

Erosion

The early settlers of the United States grew cotton, tobacco, and corn in wide rows, which they tilled to control weeds. That practice promoted erosion and loss of organic matter because the fields were partially or completely bare for much of the year. When the soils lost productivity, the farmers moved on, clearing and plowing new fields. They moved west, settling more and more of the country's seemingly limitless land. By the end of the 19th century, the westward movement had climaxed, and people had begun to realize that they must conserve soil resources.

Data from surveys conducted in the middle 1930s show the devastating effects of mismanagement of U. S. farmland: 24% of the existing farmland had lost half or more of its topsoil, 12% was essentially ruined, and 12% was severely damaged. Another 24% of the farmland showed measurable erosion. Only 28% of the farmland remained unaffected (Troeh et al., 1991). This situation left large areas of bare soils, which were vulnerable to wind erosion.

In May 1934, a dust storm that originated in the southern Great Plains carried an estimated 200 million tons of soil to the northeastern United States and out to sea. Approximately one year later, as Hugh Hammond Bennett made his plea to Congress to pass a bill that would create a permanent Soil Conservation Service, another major dust storm blew into Washington, D.C. (McMillen, 1981). The bill passed. Since the mid 1930s, U.S. farmers have made great progress controlling both wind and water erosion — commonly with the support of governmental programs. However, erosion remains a serious problem.

- ***Lack of vegetation***
- ***Water flowing across bare soils***
- ***Wind blowing across bare dry soils***
- ***Formation of salts, crusts, and physical or chemical layers***
- ***Pollution from radioactive fallout***
- ***Pollution from organic contaminants***
- ***Contamination with cadmium, mercury, and other heavy metals***
- ***Sinking of organic soils due to volume loss from oxidation***
- ***Disturbances from urban development and other activities***

Wind is a major erosive agent only when the soil surface is bare and dry (Fig. 23). Sandy soils are especially subject to wind erosion because particles do not stick together. Wind also erodes organic soils (such as peats and mucks) and soils from volcanic ash, when they are dry, because the dry particles are very light and noncohesive.

Overgrazing causes soils to lose their cover of plants. When that occurs, soil degradation is inevitable. The impact of rain drops and water flowing over bare soil releases individual particles and destroys soil aggregates. Because wind and water can move individual soil particles quite easily, erosion is accelerated.

In humid regions, water is a much more visible and effective agent of erosion than wind, especially on sloping lands (Fig. 24). Water running over bare soils or in streams with exposed unconsolidated materials along the banks is almost always muddy. A well-rooted vegetative cover, such as permanent sod that slows surface flow, is good protection against erosion because the roots prevent water from detaching soil particles. But, such cover is commonly absent in conventionally tilled cropland and during developmentof urban areas.

Fig. 23. *Severe erosion by wind can result in dust storms like the one pictured here in Burkina Faso, West Africa (Tropical Soils Research Program, Soil & Crop Sci. Dept., Texas A&M Univ., College Station, TX).*

Fig. 24. *Most soil loss from erosion by water is the result of sheet and rill erosion. They are less apparent than gully erosion, however.*

Dust storms that plagued the Great Plains of the United States during the 1930s focused national attention on soil erosion. Because of the size and severity of these storms, the region came to be called the "Dust Bowl," and the time period, the "dirty thirties." The heart of the Dust Bowl consisted of nearly 100 million acres in the panhandles of Texas and Oklahoma as well as adjacent parts of Colorado, New Mexico, and Kansas. To a lesser extent, dust storms were also a problem over much of the Great Plains, from North Dakota to west central Texas.

On those hot, dry days, blowing soil was everywhere. "Black blizzards"as blinding as a snowstorm reduced visibility to a few feet and mired buildings and machinery in great dunes of windblown soil. In the spring of 1934, a wind storm that lasted for a day-and-a-half created a dust cloud that extended for 2000 kilometers (1200 miles). As the sediment moved east, "muddy rains" were experienced in New York, and "black snows," in Vermont. Less than a year later, another storm carried dust more than 3 kilometers (2 miles) into the atmosphere and transported it 3000 kilometers from its source in Colorado to create twilight conditions in the middle of the day in parts of New England and New York.

The expansion of agriculture set the stage for the dust bowl. Mechanization allowed the rapid transformation of the grass-covered prairies of this semiarid region into farms. Between the 1870s and 1930, the area of cultivation in the region expanded nearly tenfold, from about 10 million acres to more than 100 million acres. As long as precipitation was adequate, the soil remained in place. Without moisture, grasses and plants that normally held the soil in place withered and died, leaving the unprotected fields vulnerable to the wind. The resulting soil loss, crop failures, and economic hardship led to the establishment of the Soil Conservation Service in 1935 within the U.S. Department of Agriculture.

Since the 1930s, soil conservation practices have significantly reduced the rate of wind and water erosion on U. S. croplands. These practices include terraces, residue management, cover crops, crop rotations, water and sediment control basins, field and contour strip cropping, herbaceous wind barriers, and filter strips. However, changes in the weather, such as high rainfall or heavy winds, make the fight to reduce erosion a continuing challenge that impacts society and the environment.

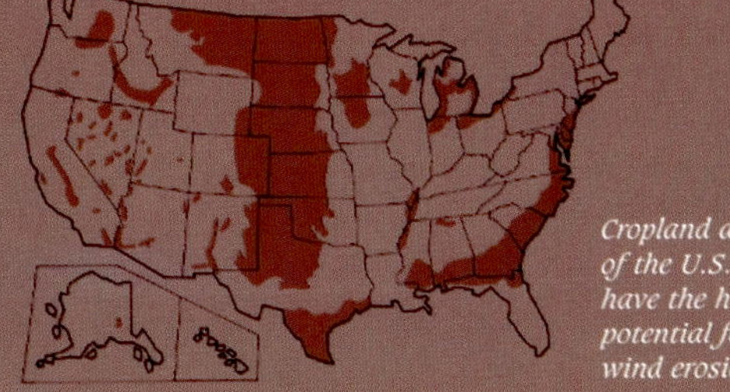

Cropland areas of the U.S. that have the highest potential for wind erosion.

Fig. 25. *Views of the North Sulphur River in Fannin County, TX. Before 1930, the former river bed could be seen from Highway 50. Today, as a result of engineered channeling, the river bed is approx. 80 m (250 ft) wide. Although it never floods the former flood plain, the river has tremendous erosive impact wherever tributaries enter because of the velocity of water flow and the channel depth to which tributaries tend to cut.*

The erosiveness of moving water is a function of the velocity of flow. The steepness and length of the slope along which water flows determine velocity. As slope steepness increases, the pathways where water flows require special protective measures, such as slowing or trapping surface flow. In much of North America, intense rain storms of short duration commonly occur, and many soils cannot absorb all of the rainfall. Thus, surface runoff is inevitable and land even on modest slopes erodes when it is left uncovered.

Some clayey soils shrink when they are dry and swell when they are wet (Fig. 14). The cracks that form as soils dry provide pathways for rapid entry of water when wetting occurs. Although this fast absorption increases the rate of water intake and decreases surface runoff, it tends to destabilize the banks of streams and gullies (Fig. 25). Soils on steep slopes may slump and collapse downslope causing landslides, especially when vegetation is sparse or absent and the soil is wet (Fig. 26). Built-up areas on the steep hill and mountain slopes of California are subject to just such soil instability.

Fig. 26. *Hillslope slump near Eugene, OR. Partial clearing for home and road construction has destabilized soil on steep slopes.*

Depletion of Organic Matter and Nutrients

Soil organic matter is an energy source for organisms that have an important role recycling nutrients for plants. In addition, organic matter is a nutrient source, nutrient retainer, and a contributor to structural stability in soil. If soil quality is to be sustained, it is necessary to maintain a healthy environment for soil organisms and for crop plants.

Organic matter and nutrients are commonly reduced in soils that have been mismanaged. The depletion of organic matter and nutrients results in lower soil productivity. It is difficult to manage and control organic matter in soils. On one hand, the breakdown of organic matter is desirable because it releases nutrients for plants. On the other hand, the breakdown

How America's farmers and ranchers use and manage their land is key to producing the nontraditional agricultural commondities that people value and to maintaining healthy, stable landscapes and watersheds.

— *America's Private Land, A Geography of Hope*

and loss of organic matter is undesirable because the soil loses its water-holding capacity and good structure. Achieving a balance between inputs and losses of organic matter is most desirable. Over-working of soil or excessive drainage will introduce too much oxygen and encourage the breakdown of organic matter. Returning plant and animal residues to the soil and reducing erosion will increase the levels of organic matter. Long-term stability depends on maintaining a proper balance between inputs and losses of organic matter.

Loss of Biodiversity

One measure of soil quality is biodiversity. According to ecologists, highly diverse communities, in which soil organisms are evenly distributed among a large number of species, generally characterize stable soil ecosystems. Several species handle each task and none becomes dominant.

Naturally occurring organic materials are the foodstuff for much of the biological community in soils. Most micro- and macro-organisms require an external source of organic carbon. The level of, and to a great extent, the diversity of organismal activity is a function of the kinds and amounts of organic substrates that are available. Organic compounds that will adversely affect soil organisms may be introduced. For example, nematicides are used with intent to kill soil nematodes. Other pesticides that are used to control weeds, insects, or diseases enter the soil where they may kill other plants and animals, or worse yet, they may move through the soil in drainage water thereby contaminating ground and/or surface waters.

When nutrients, such as nitrates and phosphates, move directly into lakes and rivers, they cause excessive growth of aquatic plants and algae. As the oxygen supply in water dwindles, fish and other aquatic life die. The main sources of phosphates and nitrates are treated sewage and runoff from farms and urban areas. Nitrates are supplied in limited quantities by decaying plant material and nitrogen-fixing bacteria, but phosphates must come from bones, organic matter, or phosphate rocks. Consequently, the use of phosphate detergents, combined with agricultural or urban runoff, has greatly affected many water bodies. Some have become ecological disasters.

Virtually all types of land use change the ecosystems in soils. For example, growing a single crop or turfgrass almost inevitably decreases biodiversity among organisms in the soil beneath it. Adding degradable organic materials and nutrients will improve a soil, increase the level of activity among soil organisms, and perhaps even increase their diversity. Many cultural practices affect soil quality, and maintaining biodiversity is an important consideration in determining how to manage soils.

ENVIRONMENTAL CHALLENGES IN SOIL CONSERVATION

- **Managing cultivation to maximize production of food and fiber and minimize erosion.**
- **Managing irrigation to maximize soil productivity and water-use efficiency while minimizing salt contamination.**
- **Maintaining a balance between inputs and losses of organic matter in soil.**

Contamination

Contamination of soil and water resources has been a byproduct of gathering, transporting, processing, storing, using, and disposing of the chemicals on which modern society depends. The use and disposal of consumer products can result in contamination when fertilizers, pesticides, heavy metals from sewage sludge, and solvents from cleaning fluids and paints accumulate in the soil or leach into the groundwater. Some contamination results from deliberate activities that lead to excess accumulation of pesticides, wastes, or other contaminants in the soil. Contamination can also be a result of an accidental spill during oil production or the transport or storage of fuels or chemicals. Some of the sources of contamination, typical contaminants, and the resultant problems are listed in Table 2 and discussed in the following section.

Sources of Soil Contamination

Source	Contaminants	Resulting problem
NATURALLY OCCURRING:		
Salt seeps	Chloride and sulfate salts of sodium, calcium, and magnesium	Poor or no plant growth, erosion
Metals	Molybdenum	Toxicity to cattle
MINING AND INDUSTRIAL ACTIVITIES:		
Acid-mine drainage	Strong acids	Poor or no plant growth, erosion
Metal-contaminated dust deposition	Lead	Toxicity to children
Petroleum and chemical spills	Oil, gasoline, solvents	Dead vegetation
Acid rain	Strong acids	Poor plant growth, acid runoff
Salt spills	Sodium chloride	Poor or no plant growth, erosion
Erosion from waste tailing and piles	Strong acids, metals	Poor or no plant growth, erosion
Storage-tank leaks	Gasoline, solvents	Groundwater pollution
Pipeline leaks	Oil, gasoline, solvents	Dead vegetation
Transportation spills and leaks	Fuels, organic chemicals, acids	Soil and groundwater pollution
AGRICULTURAL ACTIVITIES:		
Biocides	Spilled or over-applied pesticides	Dead vegetation
Fertilizers	Nitrate	Toxicity to cattle, groundwater pollution
Animal feeding (feedlots)	Nitrate, ammonium	Groundwater contamination
Irrigation with saline water	Sodium chloride	Poor or no plant growth, erosion
WASTE DISPOSAL:		
Sewage-sludge application	Cadmium	Plant uptake, toxic to animals
Oily-waste application	Oil	Several years before plants will grow
OTHER:		
Military	Lead, TNT	Toxicity, soil contamination
Radioactive fallout	Strontium	Toxicity, soil contamination

Table 2

Salt Contamination. Soils can become contaminated with salt as a result of natural salt seeps, spills, or leaks of salt solutions or brines; irrigation with saline water; runoff from salt-storage facilities; and as a result of roadway or sidewalk de-icing. Excess salts in the topsoil will decrease plant growth and at high concentrations can completely prevent it. Salts are readily leached from permeable sandy soils, but they are more difficult to leach from fine-textured clay soils. Adequate drainage must be available to allow the salts to flow deep into the soil so that they will not accumulate in the root zone.

Irrigation without proper drainage to remove salts and prevent a rise of the water table can change soils from being highly productive to unusable. Irrigation and drainage waters, manures and wastes, sewage sludge, mineral fertilizers, and industrial brines are sources of salt. In situations where irrigation water is salty and deep drainage is too slow, it is necessary to install slotted pipes across fields at depths of about 1.8 m (6 ft) to allow saline drainage to flow into streams.

Using the best technology, irrigation may be sustainable. Unfortunately, short-term economic incentives, inadequate knowledge of best management practices, water-rights laws, and adverse effects of drainage waters on the environment often discourage the adoption of sound irrigation practices. Approximately one-third of the global harvest comes from the 16% of cropland that is irrigated (Pastel, 1994). The amount of irrigated land increased about 1.4% per year from 1978 to 1991. That increase was slower than the rate of population growth and slower than the rate that irrigated land increased between 1961 and 1978. Thus, the amount of irrigated land per person is decreasing. Excessive salt accumulation in previously irrigated land and rapid population growth are factors affecting this trend.

Acid Rain. In areas where automobiles are concentrated or high sulfur coal is burned as a source of energy, the air becomes contaminated with acid-forming chemicals. When rain deposits them on the soil, they acidify it. The increase in acidity causes essential nutrients, including calcium and magnesium, to leach from the root zone and results in decreased growth of vegetation.

Mine Lands. Mining activities, including both underground and strip mining for coal, metals, and aggregates including sand and gravel, have resulted in large areas being covered with mine wastes. Many of these areas were originally left to revegetate naturally without being leveled. Among the contaminants that inhibit the growth of vegetation and future land use are pyrite, which releases iron, other metals, and sulfuric acid when exposed to the air. Federal regulations now require that the leftover materials from all coal mines be leveled and revegetated. Each mine site poses a unique set of conditions that must be overcome to allow plants to grow, so the process of reforming a surface soil can begin.

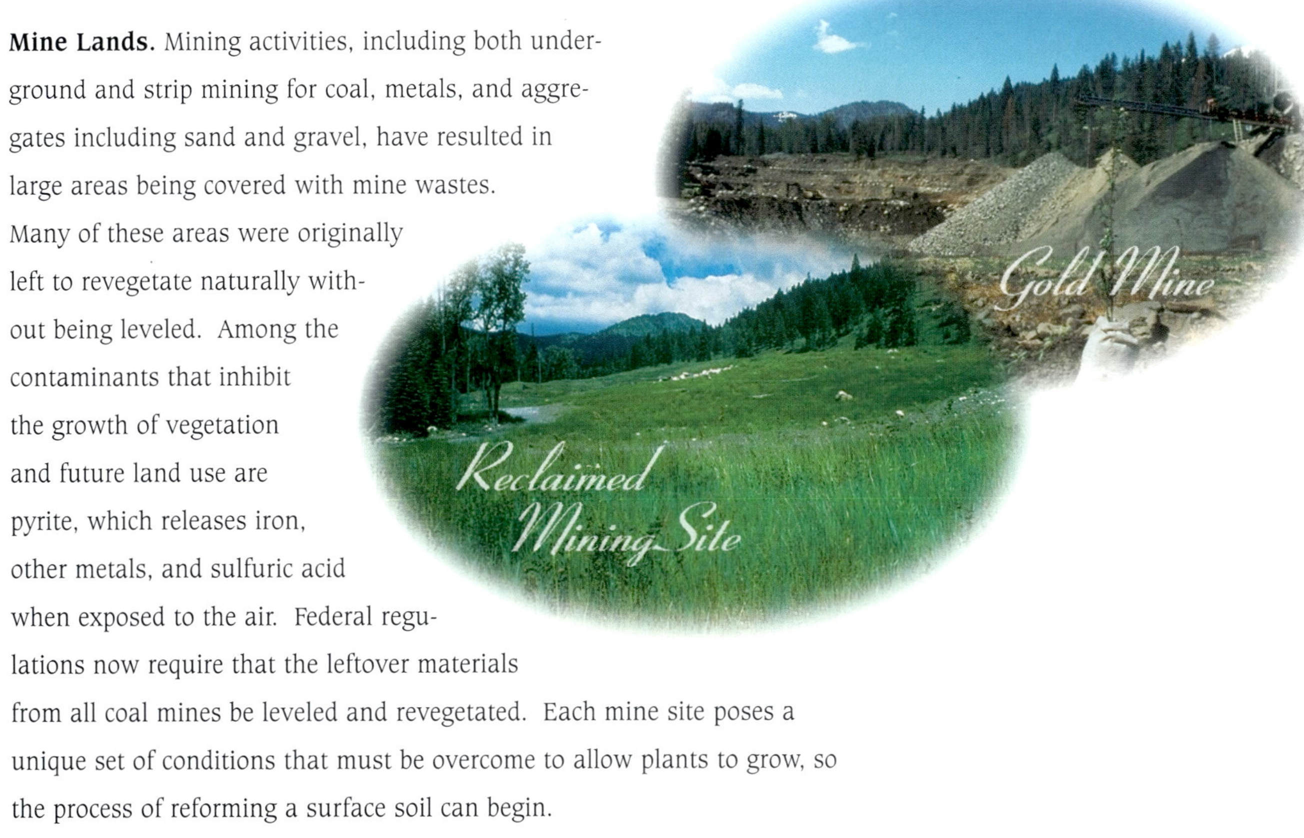

Heavy Metals. At low concentrations, heavy metals (such as cadmium, lead, or zinc) are not toxic to plants and animals. In fact, small amounts of many metallic elements are essential for proper nutrition. However, at elevated concentrations, metals are toxic to plants and animals. The addition of large quantities of sewage sludge can serve as a source of excess heavy metals in the soil, as can wind-blown dust from mine tailings, smelters, and metal processors. In some areas, metals such as lead or arsenic have accumulated to concentrations toxic to residents, and in extreme cases, soils have become so contaminated that vegetation will not grow. Modern air-quality standards have greatly diminished metallic air pollutants, and many areas that have been contaminated are now undergoing remediation.

The Environmental Protection Agency does not allow sewage sludges with excessively high metal contents to be spread on the soil. Those with lower concentrations can be safely spread on the land as long as the cumulative metal concentrations do not exceed health-based standards. Table 3 shows typical concentrations of some heavy metals in native soil as well as the maximum concentrations allowable in soils that have been amended with sewage sludge. Since the top 15 cm of soil typically weighs about 2 million kilograms per hectare (about 2 million pounds per acre for a 6-inch depth), sewage can sometimes be applied at rates that would supply the nutrients needed by crops for many decades before metal accumulation would be a problem.

Groundwater supplies about 20% of all water used in the United States. The Environmental Protection Agency has established a contaminant level of 10 mg nitrate-nitrogen per liter (or 45 mg nitrate per liter) as the maximum safe level for drinking water.

Concentrations of Some Heavy Metals in Soil

Element	Typical concentration of metals in native soil[b] (milligrams per kilogram)	Maximum allowable metal concentration in soils amended with sewage sludge[a] (milligrams per kilogram)
As (Arsenic)	5.2	41
Cd (Cadmium)	0.06[c]	26
Cr (Chromium)	37	1,500
Cu (Copper)	17	750
Hg (Mercury)	0.058	17
Ni (Nickel)	13	210
Pb (Lead)	16	300
Se (Selenium)	0.26	100[d]
Zn (Zinc)	48	1,400

[a] Federal Register 58:(32):9248-9380, Friday Feb. 19,1993. 40CFR Parts 257, 403, and 503. Standards for the Use or Disposal of Sewage Sludge.
[b] Source: U.S. Geological Survey, 1992.
[c] Source: Brown, 1983.
[d] Except for alkaline soils

Table 3

Spills and Leaks. Spills and leaks occur during the production, storage, transport, use, and disposal of products and chemicals. They most frequently occur at oil and gas wells, industrial sites, military facilities, gasoline stations, and along pipelines, railroads, and highways. If the product or chemical that spills is a liquid, it will soak into and contaminate the soil. Soluble salts spilled on the soil surface will dissolve in rain water and leach into the soil. Solids and dust contaminated with metals or other insoluble chemicals can be slowly incorporated into the soil by worms and burrowing insects or as a result of cultivation.

Spilled material may be sufficient to prevent plant growth, thus opening the soil to erosion. If the ability of the soil to adsorb the spilled materials is exceeded, they may drain through the soil and contaminate groundwater.

Agricultural Activities. Excessive applications of fertilizers and pesticides, particularly herbicides that are intended to kill weeds, contaminate soil. Such instances of soil contamination are typically unintentional because of the high cost of agricultural additives. Excessive nutrients and pesticides may not be harmful to vegetation, but they may drain through the soil and contaminate groundwater (Fig. 27). The possibility of contaminants reaching groundwater aquifers is a grave environmental concern because it may be practically impossible to purify them.

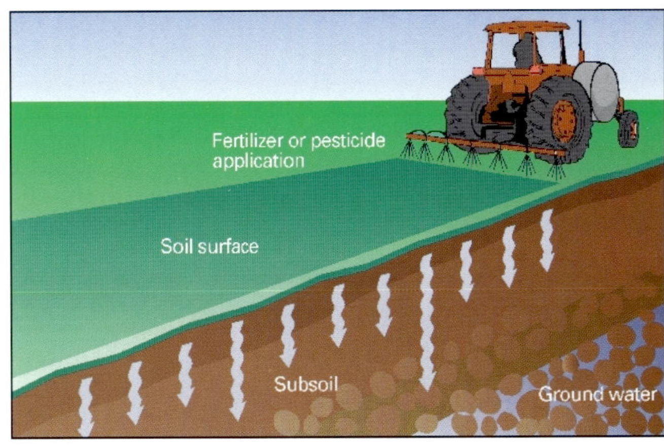

Fig. 27. *Movement of surface-applied fertilizers and pesticides through soil.*

Modern pesticides are designed to degrade in the soil shortly after their application to minimize the risk of contamination and leaching into the groundwater. However, even with such precautions, a small fraction of some pesticides applied to soil can leach to the groundwater, particularly through sandy soils. Fertilizers and pesticides also may leak into groundwater from accidental spills when streams or poorly sealed wells are located nearby.

Waste Disposal. Soils have the capacity to degrade large amounts of organic waste. In a natural ecosystem, many tons of dead vegetation, animals, and insects are degraded annually in a hectare (2.47 acres) of topsoil. Humans have used the soil to dispose of crop residue, animal manure, and human waste for centuries. More recently, industrial wastes such as sludges from food- and fiber-processing plants have also been incorporated into soil. These practices, if properly executed, are safe and effective. They can be done routinely without contaminating the soil or the groundwater, as long as the waste applications are managed so that the nutrients can be absorbed by plants, and as long as the waste materials do not contain excessive concentrations of salts or toxic metals that accumulate in soil.

Radionuclides. Soils have become contaminated with radioactive fallout resulting from the atmospheric detonation of nuclear bombs; nuclear accidents such as the one that occurred at Chernobyl, in 1986; dust from mine tailings; past waste-disposal activities; and accidental spills. Airborne radionuclides can travel great distances, and may be deposited thousands of miles from their source, thus causing global contamination. People are exposed to radionuclides by consuming contaminated water, plants, or animals and by dust that is resuspended from contaminated soils. Problems resulting from this exposure include an increased risk of cancer and birth defects.

Industrial and Military Activities. Improper disposal of wastes from industrial and military sites has contaminated soils in some areas. Military activities including bombing and rifle ranges, as well as the manufacturing of munitions and toxic chemicals also have resulted in soil contamination. The contaminants include lead, trinitrotoluene (TNT), pesticides, and volatile organic solvents. An example of a highly-toxic location is the Rocky Mountain Arsenal near Denver, Colo. Many of these military sites are now being restored, and many industries have improved their waste disposal practices to minimize soil contamination.

Soil as a Contaminant. Soil itself can be a contaminant when it is eroded into streams and lakes or when wind-blown dust contaminates the air. Erosion is a natural process. It is held in check by vegetation growing on the soil surface. Even with our best management practices, however, some quantity of topsoil typically erodes each year even from a pasture maintained in permanent grass. When the vegetation is removed, as is often necessary to produce crops, wind and water can carry large amounts of soil from a field. Soil washed into streams and ponds makes them muddy, limiting the growth of desirable plants and fish. When soil accumulates as sediment in slow-flowing streams and behind dams, it can plug the streams so that the reservoirs behind the dams become useless for water storage.

Erosion rate by wind on U.S. croplands decreased by 25% in 1982-92.

Erosion rate by water on U.S. croplands decreased by 24% in 1982-92.

Improving and Sustaining Our Soils

I conceive that land belongs to a vast family of which many are dead, few are living, and countless numbers are unborn....

— *Nigerian Chieftain*

Pesticides by their very nature are toxic. They become pollutants when they appear in groundwater, accumulate in the food chain, or reside as residues on food. Pesticides that break down quickly in the environment minimize exposure to nontarget organisms.

Within the past few decades, laws have been passed and massive efforts have been undertaken to clean up contaminated soils and to prevent further contamination. Because past contamination is extensive and accidental spills will continue to occur, there is much interest in developing economically and ecologically sound methods of treating contaminated soils.

Reclaiming Damaged Soils

The nature of the contaminants and their chemical properties determine which method or methods should be used to reclaim damaged soils. Typical procedures include leaching, biodegradation, composting, vapor extraction, vegetative remediation, and soil removal.

Leaching. Leaching is accomplished by flushing excess water through soil to remove water-soluble contaminants, including salts and nutrients. Its success depends on the soil being maintained in a sufficiently permeable state and the concentration of contaminant in the leachate being sufficiently low to avoid pollution of groundwater.

Biodegradation. Biodegradation is the process by which microorganisms, which are usually already present in the soil, degrade organic contaminants such as oil, sewage sludge, sawdust, or spilled molasses. The process takes place without intervention. However, the rate of degradation may be speeded by plowing to mix the contaminants with the soil, adding fertilizer to supply deficient nutrients, adding organic matter such as manure or old hay to stimulate microbial action, aerating the soil, or irrigating it in dry climates. Crude oils in warm, moist soils typically degrade to one-half their initial concentration during the first year, and the residual decreases in half again each succeeding year. Easily degraded contaminants such as molasses will degrade in the soil in a matter of weeks.

Composting. Soil contaminated with TNT, spilled pesticides, and other slow-to-degrade chemicals may be excavated, mixed with a large amount of easily degradable organic matter such as manure, and placed in piles to compost. During composting, the naturally occurring microorganisms flourish. Their activity raises the temperature of the compost pile to

over 50°C for several days and degrades the organic waste as well as the target compounds.

Vapor Extraction. Soils contaminated with volatile organic chemicals, such as benzene from gasoline or trichloroethylene from a spilled solvent, may be excavated and placed under plastic on an impermeable surface over perforated pipes (Fig. 28). Air is drawn through the soil into the pipes by a pump that develops a vacuum, thus removing gaseous contaminants from the soil. The gases may be caught on a carbon trap, or may be burned in an incinerator, to prevent air pollution. When volatile contaminants have migrated deep into the soil, vapor extraction can be done in place through wells installed into the unsaturated soil.

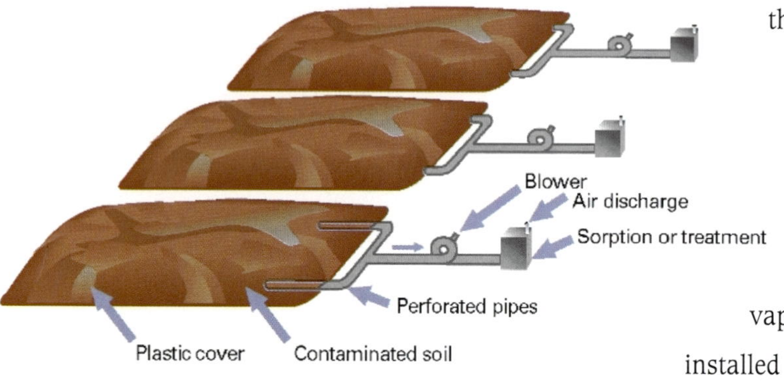

Fig. 28. *Vapor extraction system used to remediate fuel or solvent-contaminated soils.*

Vegetative Remediation. Plants may take up excess nutrients such as nitrate, thus minimizing the potential for groundwater pollution. Selected plants may also be used to remove excessive levels of salts and certain toxic metals from soils. Only a few plant species, including members of the mustard family, can take up enough metal to be useful in this regard, and none of these plant species are commonly grown crops.

Soil Removal. When the soil is polluted with excessive amounts of radionuclides, toxic metals, or concentrated pesticides from a spill, there may be no choice but to dig up the soil and dispose of it in a landfill. If removal can be organized quickly after contamination occurs, only the top centimeter or so of the soil may need to be removed. If several years pass, or if the soil is cultivated, it may be necessary to remove the top 15 to 30 cm (6 to 12 inches), to remove metals or radionuclides. For spilled pesticides or other liquid chemicals, it may be necessary to excavate to much greater depths. Contaminants that accumulate and are slow to move in soil do not present major hazards in modern landfills because the contaminants are very slow to migrate from these facilities.

Sustainability, Soils, and Society

Sustainability simply means that resources should be used to provide for the needs of the present generation without compromising the ability of future generations to meet their own needs. Maintaining soil quality and productivity over time means different things to different people. To some, sustainability implies encouraging a return to natural systems and discouraging the use of mineral fertilizers and pesticides. Others contend that increases in productivity can, in large measure, be attributed to development of responsive plant varieties and inputs of fertilizers, soil amendments, and pesticides. Both groups agree that the organic component of soils is crucial to overall soil quality. Because soil quality decreases with use, we should encourage agricultural practices that provide both high productivity and good soil quality over long periods of time.

In much of the developing world, population pressures, lack of resources to make recommended soil improvements, lack of access to existing technology, and economic conditions combine to discourage practices that will maintain soil quality. Furthermore, in many humid tropical settings, ecosystems are much more fragile than in temperate zones. The ecosystems are fragile because a very high percentage of the nutrients are in organic forms as opposed to the more stable mineral forms of less highly weathered soils. Badly eroded and eroding landscapes and saline soils can be observed in almost all habitable areas of the world. The evidence shows that sustainability has not been achieved for much of the world's land area.

Soil degradation leads to declines in productivity. Using existing knowledge and expertise, degradation can be reversed on a broad scale. Degraded soils can be reclaimed; irrigation water with lower salt content can be used; watersheds can be managed better; nutrients can be added and used more efficiently; and polluted soils can be remediated. Thus, management practices must minimize erosion, while protecting surface and groundwater from contamination by salts, nutrients, and pesticides.

Some of our most productive soils are found in regions where the soils are naturally fertile, where the climate or irrigation provides adequate moisture during a warm growing season, and where winters have temperatures low enough to suppress weeds, insects, diseases, and to preserve soil organic residues. Protection of our most productive soils will require a shift in the way we think about land use and value. Either the price of food and fiber will increase enough to allow productive farm land to be more valuable for growing crops than for development for housing, or additional laws will need to be passed to protect productive soils from urban expansion.

Earth systems are complex and dynamic. Natural processes and human activities affect these systems and the environment — soil, water, air, plants, and animals. Because soils are the very foundation of our existence, the more we understand the workings of Earth processes and systems, the better chance we have of sustaining our soils and society.

STEPS TOWARD SUSTAINABILITY

Develop economically and ecologically sound methods to reclaim damaged soils and increase agricultural productivity

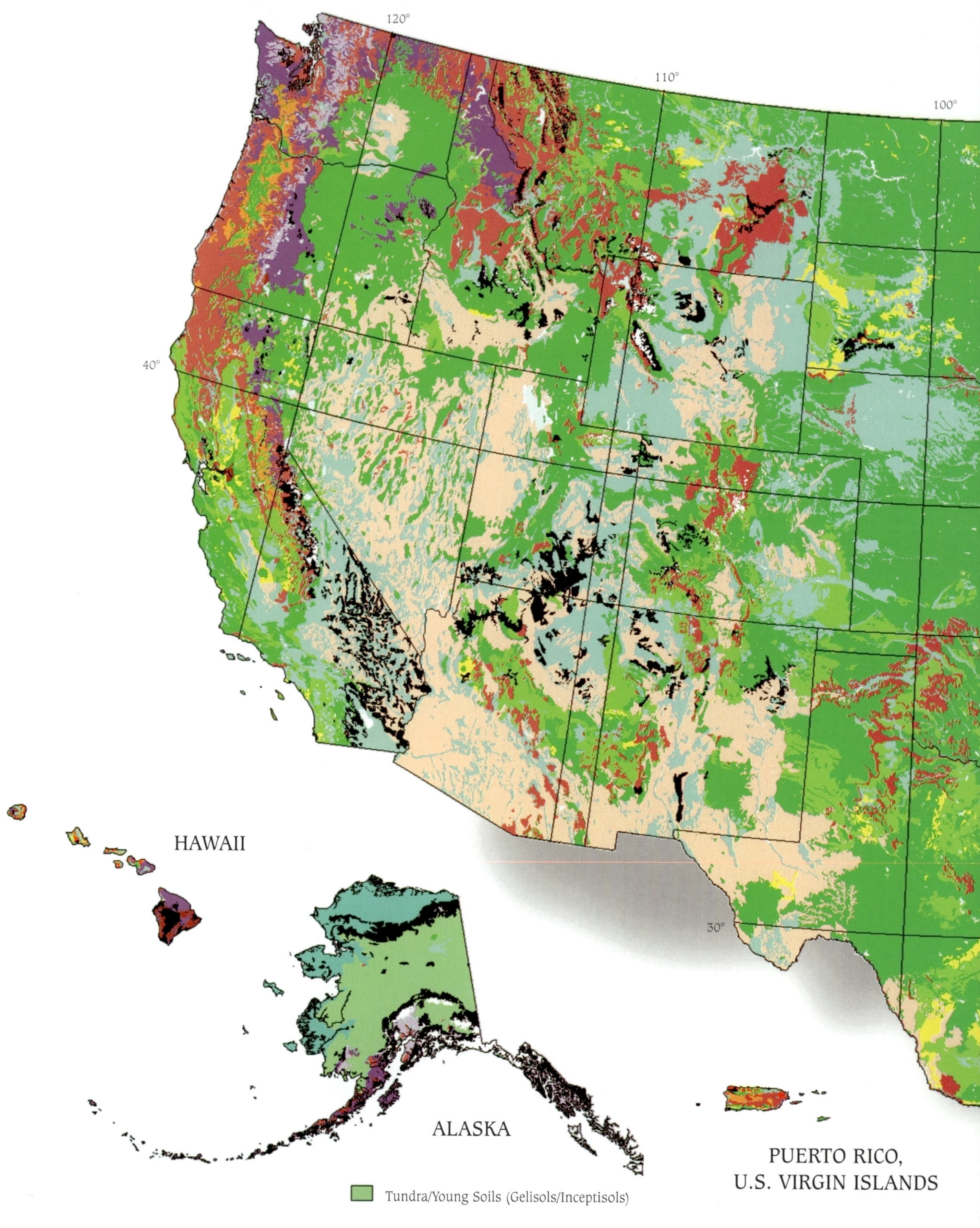

120°
110°
100°
40°
30°
HAWAII
ALASKA
PUERTO RICO,
U.S. VIRGIN ISLANDS
Tundra/Young Soils (Gelisols/Inceptisols)

Geographic Distribution of Soils

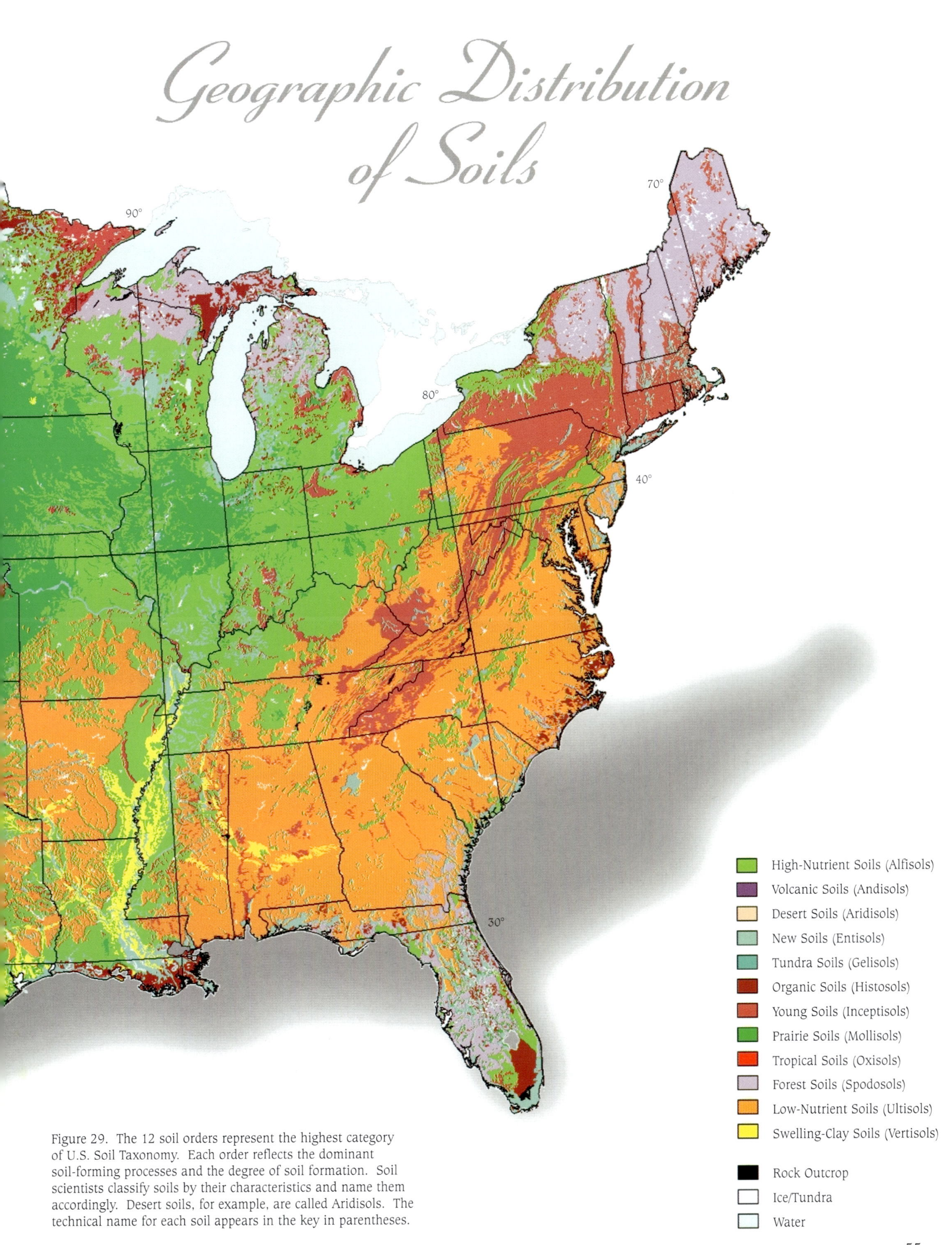

Figure 29. The 12 soil orders represent the highest category of U.S. Soil Taxonomy. Each order reflects the dominant soil-forming processes and the degree of soil formation. Soil scientists classify soils by their characteristics and name them accordingly. Desert soils, for example, are called Aridisols. The technical name for each soil appears in the key in parentheses.

References

Bailey, Ronald (editor). 1995. *The True State of the Planet.* A project of the Competitive Enterprise Institute. The Free Press, A Division of Simon & Schuster, Inc., New York.

Bouma, J. (and others). 1972. *Soil Absorption Of Septic Tank Effluent.* Inform. Circ. 20, Univ. of Wisc. Ext., Madison, WI.

Brady, Nyle C. and Ray R. Weil. 1999, 12th ed. *The Nature and Properties of Soils.* Prentice Hall, Upper Saddle River, NJ.

Brown, K.W. 1983. *Hazardous Waste Land Treatment.* U.S. EPA Publ. SW-874.

Brown, Lester R. and Hal Kane. 1994. *Full House - Reassessing the Earth's Population Carrying Capacity.* W.W. Norton and Company, NY.

Chaney, R.L. 1980. *Health Risks Associated with Toxic Metals in Municipal Sludge.* pp. 59-83 In G. Bitton, B. Damron, G. Edds and J. Davidson (eds.). *Sludge - Health Risks of Land Application.* Ann Arbor Science Publ., Ann Arbor, MI.

Das, B.M. 1994. *Principles of Geotechnical Engineering.* PWS Publishing Co., Boston.

Foth, Henry. 1990, 8th ed. *Fundamentals of Soil Science.* John Wiley and Sons, New York.

Freedman, B. 1995. *Environmental Ecology.* Academic Press, San Diego, CA.

Hillel, Daniel J. 1991. *Out of the Earth - Civilization and the Life of the Soil.* The Free Press, A Division of Simon & Schuster, Inc., New York.

Lovelock, J.E. 1979. *Gaia—A New Look At Life On Earth.* Oxford University Press, New York.

Mausbach, M.J. 1994. *Classification of Wetland Soils for Wetland Identification.* Soil Survey Horizons 35(1):17-25.

McMillen, Wheeler. 1981. *Feeding Multitudes - A History of How Farmers Made America Rich.* The Interstate Printers and Publishers, Inc., Danville, IL.

Pastel, Sandra. 1994. *Irrigation Expansion Slowing In Vital Signs 1994 - The Trends That Are Shaping Our Future.* p. 44. Lester R. Brown, Hal Kane and David Malin Roodman. Worldwatch Institute.

Pierzynski, G.M., J. T. Sims and G.F. Vance. 1994. *Soils And Environmental Quality.* Lewis Publishers, Boca Raton, FL.

Rebuffoni, D. 1992. "Owners victory over land-use rules: $58,600 per acre." In *Star Tribune,* March 24, Minneapolis, MN.

Singer, M.J. and D.N. Munns. 1999, 4th ed. *Soils—An Introduction.* Prentice Hall, Upper Saddle River, NJ.

Trefil, James. 1997. "Nitrogen." In *Smithsonian,* Vol. 28, No. 7, pp. 70-78.

Troeh, Frederick R., J. Arthur Hobbs and Roy L. Donahue. 1999, 3rd ed. *Soil and Water Conservation.* Prentice Hall, Upper Saddle River, NJ.

U.S. Department of Agriculture, Natural Resources Conservation Service. 1996. *America's Private Land, A Geography of Hope.* Program Aid 1548.

U.S. Department of Agriculture, Natural Resources Conservation Service. 11-14-98. *National Resources Inventory - 1997 State of the Land Update.* http://www.nhq.nrcs.usda.gov/NRI/1997update/

U.S. Department of Agriculture, Soil Conservation Service. 1994. *Soil Erosion by Water.* Agricultural Information Bulletin 513.

U.S. Department of Agriculture, Soil Conservation Service. 1994. *Soil Erosion by Wind.* Agricultural Information Bulletin 555.

U.S. Environmental Protection Agency. 1988. *America's Wetlands: Our Vital Link Between Land And Water.* OPA-87-016, U.S. Gov. Print. Office, Washington, D.C.

U.S. Geological Survey. 1992. *Element Concentrations in Soils and Other Surficial Materials of the Conterminous United States.* U.S. Geological Survey Professional Paper 1270. U.S. Government Printing Office 1992-673-049/66008.

Credits

Front Cover — Hands with soil (T. Loynachan, Iowa State Univ.), Prairie soils (Mollisols) Left to right: Kansas/ Nebraska wheat field (S. Waltman, USDA-NRCS), Rangeland near Scottsbluff, Nebraska (W. Waltman, USDA-NRCS), and Agricultural/urban interface, Richmond, Virginia (NRCS photo by T. McCab); Forest soil (Spodosol) Waterfall in forest near Portland, Oregon (G. James); Tropical soil (Oxisol) Hawaii (University of Nebraska Press, A. Aandahl); Organic soil (Histosol) Vegetable crops in a former Michigan wetland (The Soil Classifiers of Michigan); Desert soil (Aridisol) Utah butte in desert landscape (S. Waltman, USDA-NRCS); Tundra soil (Gelisol) Patterned ground and tundra near an arctic lake (W. Lynn, USDA-NRCS)

Title page — Hands with soil (T. Loynachan, Iowa State Univ.)

Page 6 — Desert soil (Aridisol) Utah butte in desert landscape (S. Waltman, USDA-NRCS); Soil Profile (National Cooperative Soil Survey)

Page 7 — Figure 1, Nicollet loam soil (B. Elbert, Iowa State Univ.)

Page 8 — Table 1, Grain-Size Scale (adapted from AGI Data Sheet 29.2)

Page 9 — Figure 2, Air circulation affects color in subsoils (T. Cooper, Univ. of Minnesota)

Page 10 — Figure 3, Forest soil profile (The Soil Classifiers of Michigan)

Page 11 — Mount St. Helens (L. Topinka, USGS/Cascades Volcano Observatory)

Page 14 — Organic soil (Histosol) Vegetable crops in a former Michigan wetland (The Soil Classifiers of Michigan); Soil Profile (Florida Association of Professional Soil Classifiers)

Page 15 — Figure 4, Many roles of soil (J. De Atley, De Atley Design; T. Loynachan, Iowa State Univ.); Figure 5, Carbon Cycle (T. Hiett, Iowa State Univ.)

Page 16 — Greenhouse gases percentages (Pierzynski et al.); Figure 6, Greenhouse gases (T. Hiett, Iowa State Univ.)

Page 17 — Figure 7, Corn researcher (Ohio Ag Research and Development Center)

Page 18 — Aerial view of braided river, Resurrection River, Alaska (M. Miller, Univ. of Oregon); Mountain stream near Syria, Virginia (G. James)

Page 19 — Figure 8, Water cycle (T. Hiett, Iowa State Univ.); Figure 9, Wetlands as filters (T. Hiett, Iowa State Univ.)

Page 20 — Figure 10, Soil filters wastewater (T. Hiett, Iowa State Univ.)

Page 21 — Figure 11, Generic landfill design (T. Hiett, Iowa State Univ.); Figure 12, Drainage (T. Hiett, Iowa State Univ.)

Page 22 — Tropical soil (Oxisol) Hawaii (University of Nebraska Press, A. Aandahl); Soil Profile (National Cooperative Soil Survey)

Page 23 — Dam in northern Montana (M. Miller, Univ. or Oregon); Coarse-soil profile (Society of Soil Scientists of Southern New England, Mass.); Fine-soil profile (Professional Soil Scientists Association of Texas); Figure 13, Soil structure (T. Loynachan, Iowa State Univ.)

Page 24 — Swelling-clay soil (W. Lynn, USDA-NRCS); Figure 15, Cracks in wall (D. Williams, NRCS, retired)

Page 25 — Figure 16, Soil strength (T. Hiett, Iowa State Univ.)

Page 26 — Nutrient-poor soil (USDA-NRCS photo by T. McCab); Soil profile (Professional Soil Classifiers Assoc. of Alabama)

Page 27 — Figure 17, Total U.S. Cropland (J. De Atley, De Atley Design; Data from *National Resources Inventory - 1997 State of the Land Update,* USDA-NRCS)

Page 28 — Signage photo (H. Nuernberger, Penn State Univ.); Figure 18, U.S. Land Use, 1992 (J. De Atley, De Atley Design; Data from USDA, Soil Conservation Service, 1994)

Page 29 — Contour plowing, farmland in central Wisconsin (M. Miller, Univ. of Oregon); Mississippi River, New Orleans, Louisiana (M. Miller, Univ. of Oregon)

Page 31 — Figure 19, Strip farming (M. Milford, Texas A&M Univ.); Figure 20, "Slash and burn" cultivation (T. Loynachan, Iowa State Univ.)

Page 32 — Figure 21, Landsat™ image, Seward County, Nebraska and precision-farming equipment (W. Waltman, USDA-NRCS); Figure 22, Maize (corn) (M. Milford, Texas A&M Univ.)

Page 33 — Contour plowing (T. Loynachan, Iowa State Univ.); Irrigation, western Idaho (W. Waltman, USDA-NRCS); Corn field, St. Joseph, Missouri (S. Killgore); Text *(America's Private Land, A Geography of Hope)*

Page 34 — Erosion by water (USDA-NRCS)

Page 36 — Figure 23, Erosion by wind (Tropical Soils Research Program, Soil & Crop Science Dept., Texas A&M Univ.); Figure 24, Erosion by water, mountain stream near Syria, Virginia (G. James); sheet & rill and gully erosion (M. Milford, Texas A&M Univ.)

Page 37 — Dust Bowl photo and map (USDA-NRCS)

Page 38 — Figure 25, Views of Sulphur River (M. Milford, Texas A&M Univ.); Figure 26, Hillslope slump (M. Milford, Texas A&M Univ.)

Page 39 — Precision-farming equipment (W. Waltman, USDA-NRCS); Text *(America's Private Land, A Geography of Hope)*

Page 40 — Contour plowing, farmland in central Wisconsin (M. Miller, Univ. of Oregon)

Page 41 — Table 2, Sources of Contamination (K. Brown, Texas A&M Univ.)

Page 42 — Aerial view of salt pan/salt crystals, Death Valley, California (M. Miller, Univ. of Oregon)

Page 43 — Gold mining and reclamation sites (Gold Institute, Washington, D.C.)

Page 44 — Fault scarps and springs, Death Valley, California (M. Miller, Univ. of Oregon); Table 3, Heavy Metals (K. Brown, Texas A&M Univ.)

Page 45 — Figure 27, Fertilizer and Pesticide application (T. Hiett, Iowa State Univ.)

Page 47 — Bryce Canyon National Park, Utah (M. Miller, Univ. of Oregon); Grain (USDA-NRCS); Splash (USDA-NRCS)

Page 50 — Figure 28, Vapor extraction system (T. Hiett, Iowa State Univ.)

Page 53 — Children (USDA-NRCS)

Page 54-55 — Figure 29, Soils Distribution Map (Adapted from Dominant Soil Orders and Suborders-Soil Taxonomy 1998, United States and its territories, L. Quandt and S. Waltman, USDA-NRCS, this map has been modified through the removal of Pacific Basin islands)

All other photos not noted here are taken from the Adobe Image Library, Digital Stock Corp., or Corel Stock Photography

Glossary

acre A unit of land measure equal to 160 square rods or 43,560 square feet (.405 hectare).

adobe Sun-dried brick made by mixing soil with hay or straw.

adsorption The process by which atoms, molecules, or ions are attracted to and retained on soil surfaces.

aeration The movement of gases into and out of soil.

aerobic (a) An environment containing molecular oxygen (O_2). (b) Growing only in the presence of molecular O_2, as aerobic organisms. (c) Occurring only in the presence of molecular O_2 (said of certain chemical or biochemical processes such as aerobic decomposition).

aggregate A soil structural unit, made up of a group of individual soil particles that stick together as a unit.

agronomy The branch of agriculture that deals with the theory and practice of field crop production and soil management.

alluvium Sediments deposited on land by streams.

amendment Materials such as nutrients, lime, manure, and compost which are added to soil to increase its fertility.

anaerobic An environment lacking molecular oxygen.

anion A negatively charged ion.

available water The portion of soil water that can be absorbed by plant roots.

biocide A chemical which is used to kill any living organism.

biodegradable A substance capable of being decomposed by biological processes.

biodegradation Decomposition of organic material in the soil by microorganisms.

biomass The mass of living organisms in or on the soil.

biota Plants and animals which live in or on the soil.

biosphere The zone immediately above and below the Earth's surface which supports life.

bulk density The weight of dry soil per unit bulk volume.

carbon cycle See *soil carbon cycle.*

carbon-nitrogen ratio The weight ratio of organic carbon to nitrogen in a soil or in organic material.

cation A positively charged ion.

cation exchange The exchange of cations between a solution and those cations held on the outer surface of mineral or organic matter in the soil.

clay (a) A soil particle <0.002 mm in equivalent diameter. (b) A soil that has properties dominated by clay-size particles.

colloid A small particle that provides a large surface area per unit volume.

compost The organic residue of the natural aerobic biodegradation of organic waste which has been mixed, piled, and moistened.

degradation The process whereby a compound is transformed into simpler compounds.

ecosystem A community of organisms and the environment in which they live.

erosion The wearing away of the land surface by water or wind.

essential elements Elements required by plants to complete their life cycles.

evapotranspiration The combined loss of water by evaporation from the soil surface and by transpiration from plants.

fallow The practice of leaving land uncultivated or unplanted during at least one period when a crop would normally be grown.

fertilizer Any organic or inorganic material added to a soil to supply one or more elements essential to the growth of plants.

field capacity The soil water that remains two to three days after the soil was thoroughly wetted and after downward drainage has stopped.

frost heaving Lifting or lateral movement of soil caused by the formation of ice lenses during freezing.

GIS Geographic Information System, a computer-based system and databases that facilitate the storage, integration, and presentation of layers of data commonly on maps.

GPS Global Positioning System, a geographic and geologic system of satellites and surface equipment for determining exact locations on the Earth's surface.

greenhouse gas Gases including methane and carbon dioxide which trap heat in the Earth's atmosphere.

groundwater Water stored in saturated zones beneath the Earth's surface.

gully (a) A channel or very small valley resulting from erosion and caused by the concentrated flow of running water and through which water runs only after a rain or the melting of ice or snow. (b) Any erosion channel so deep that it cannot be crossed by a wheeled vehicle or eliminated by plowing, especially one excavated in soil on a bare slope.

heavy metals Those metals that have densities greater than 5 g per cm^3; including the metallic elements arsenic (As), cadmium (Cd), chromium (Cr), cobalt (Co), copper (Cu), iron (Fe), lead (Pb), manganese (Mn), mercury (Hg), molybdenum (Mo), nickel (Ni), and zinc (Zn).

hectare A unit of land area equaling 100 by 100 meters (1 hectare = 2.47 acres).

herbicide A chemical used for killing weeds.

horizon A layer of soil approximately parallel to the land surface and differing from adjacent layers in physical, chemical, and biological properties.

hydraulic conductivity The coefficient that determines the rate at which soil transmits liquid water through pores.

hydrologic cycle The fate of water from the time of precipitation until the water has been returned to the atmosphere by evaporation and is again ready to be precipitated.

infiltration The entry of water into soil through the surface.

irrigation The intentional application of water to the soil.

land treatment The application of wastes such as sewage sludge to soils to achieve both nutrient recycling and waste disposal.

leaching The downward movement of soluble chemicals as a result of soil drainage.

lime A soil amendment containing principally calcium carbonate and used to neutralize acidic soils.

loam A soil having properties approximately equally influenced by sand, silt, and clay components. A medium-textured soil; one with good properties for growing plants.

loess Predominantly silt-size particles transported and deposited by wind.

macronutrient A plant nutrient required in large amounts. Macronutrients include nitrogen (N), phosphorus (P), potassium (K), calcium (Ca), magnesium (Mg), and sulfur (S).

micronutrient A plant nutrient required in small amounts, including boron (B), chlorine (Cl), copper (Cu), iron (Fe), manganese (Mn), molybdenum (Mo), and zinc (Zn).

mineralization Microbial conversion of elements bound in organic forms to inorganic forms.

mottled soil A soil with a mixture of gray and reddish brown or other colors indicating seasonal waterlogging.

muck soil A soil containing 20% to 50% well-decomposed organic matter, occurs in very poorly drained locations.

nitrogen cycle The sequence of biochemical changes undergone by nitrogen wherein it is used by a living organism, transformed upon the death and decomposition of the organism, and converted ultimately to its original form.

organic soil A soil that contains greater than 20% by weight organic carbon.

parent material The mineral or unconsolidated organic deposits from which soil develops.

particle size The effective diameter of a soil particle measured by sedimentation or sieving.

peat soil A soil containing more than 50% organic matter, usually consisting of slightly decomposed to undecomposed organic residues. It occurs in very poorly drained locations.

permeability The ease with which gases, liquids, or plant roots penetrate or pass through soil.

pH A symbol for the degree of acidity or alkalinity. The negative logarithm of the hydrogen ion activity of a soil. A pH of 7 is neutral, while lower pH values indicate acid conditions and values greater than 7 indicate alkaline conditions.

plowing A tillage operation performed with a plow or disk to shatter and invert the soil.

pores The spaces between the particles in a soil.

precision farming The precise application, using GIS-GPS technology, of varying amounts of soil amendments to different locations in a field to optimize production or profit.

radionuclide An element such as uranium which emits radiation by the spontaneous disintegration of atomic nuclei.

remediation The process of correcting or reducing detrimental conditions.

runoff That portion of precipitation that flows over the land surface to a stream.

saline soil A soil containing sufficient soluble salt to adversely affect the growth of most crop plants.

sand (a) A soil particle between 0.05 and 2.0 mm in diameter. (b) A soil composed of a large fraction of sand-size particles.

saturated soil A soil in which all the pores are filled with water.

shifting cultivation A management system in which land is used one or a few years for crop production followed by a much longer period, perhaps decades, for the regrowth of native vegetation before being used for crops again.

shrink-swell clay Clay soils which undergo significant volume changes when the moisture content changes.

silt (a) A soil particle between 0.002 and 0.05 mm in diameter. (b) A soil composed of a large fraction of silt-size particles.

smectite A group of clay minerals characterized by very small particle size, large surface area, and high shrink-swell potential.

soil The unconsolidated material on the surface of the Earth that serves as a natural medium for the growth of land plants.

soil air The gaseous phase of the soil, which is that volume not occupied by solid or liquid.

soil carbon cycle The sequence of transformations whereby carbon dioxide (CO_2) is converted to organic forms by photosynthesis, recycled through the soil, and ultimately returned to its original state through respiration or combustion.

soil conservation Activity to protect the soil against physical loss by erosion or against excessive loss of fertility.

soil map A map on which the boundaries between different soils are shown. Such maps, found in soil survey reports, are available at libraries for the majority of the counties in the United States.

soil order The highest category of U.S. Soil Taxonomy. Each order reflects the dominant soil forming processes and the degree of soil formation. The 12 dominant U.S. soil orders are Alfisols (high-nutrient soils), Andisols (volcanic soils), Aridisols (desert soils), Entisols (new soils), Gelisols (tundra soils), Histosols (organic soils), Inceptisols (young soils), Mollisols (prairie soils), Oxisols (tropical soils), Spodosols (forest soils), Ultisols (low-nutrient soils), and Vertisols (swelling-clay soils).

soil sample A representative sample taken from an area, which is used to determine soil physical, chemical, and biological properties.

soil strength A measure of the resistance of a soil to shear or penetration.

soil structure The combination or arrangement of individual soil particles into larger units with a characteristic shape.

soil survey report A report that tells about the soils of an area, usually a county, and contains soil maps.

soil texture The relative proportions of sand, silt, and clay in a soil.

soil water retention The weight or volume of soil water that is retained in the soil at a given tension or suction by surface attraction.

subsidence Volume collapse caused by accelerated biological oxidation of soil organic matter when a wet soil is artificially drained.

subsoil Layers of soil below the surface soil layer.

tensiometer A device for measuring the tension with which water is held in the soil.

terrace A level, usually narrow, plain bordering a river, lake, or the sea. Rivers sometimes are bordered by several terraces at different levels.

topsoil The surface layer of soil often consisting of the A horizon that is mixed during cultivation.

transpiration The evaporation of absorbed water from plant surfaces.

weathering The physical and chemical breakdown of minerals as soil forms.

water table The upper surface of the groundwater, or saturated zone, in the soil.

wetland An area of waterlogged soils which supports the growth of cattails and other water-loving vegetation.

yield The amount of a specified crop or vegetation produced per unit of land area.

Index

Career Opportunities

Soil science graduates may find job opportunities as field soil scientists with the Natural Resources Conservation Service, state conservation agencies, private industry, or regulatory bodies. There are opportunities with consulting firms that help local and state governments deal with land reclamation and waste disposal, and that conduct environmental impact studies. With graduate work, students may work in universities, governmental agencies, industry, and consulting as soil chemists, soil microbiologists, soil mineralogists, land-use specialists, or soil physicists.

For more information:

American Geological Institute
4220 King Street
Alexandria, VA 22302
(703) 379-2480
www.agiweb.org
www.earthsciweek.org

Soil Science Society of America
677 South Segoe Road
Madison, WI 53711
(608) 273-8095
www.soils.org

USDA, Natural Resources Conservation Service/
Soil Survey Division-National Soil Survey Center
100 Centennial Mall N, Room 152
Lincoln, NE 68508
(402) 437-5499
www.nssc.nrcs.usda.gov

Soil and Water Conservation Society
7515 Northeast Ankeny Road
Ankeny, IA 50021
(515) 289-2331
www.swcs.org